Machine Learning for Radio Resource Management and Optimization in 5G and Beyond

Machine Learning for Radio Resource Management and Optimization in 5G and Beyond highlights a new line of research that uses innovative technologies and methods based on artificial intelligence/machine learning techniques to address issues and challenges related to radio resource management in 5G and 6G communication systems. This book provides comprehensive coverage of current and emerging waveform design, channel modeling, multiple access, random access, scheduling, network slicing, and resource optimization for 5G wireless networks and beyond.

This book is suitable for researchers, scholars, and industry professionals working in different fields related to mobile networks and intelligent systems. Additionally, it serves as a hands-on resource for students interested in the fields of cellular networks (5G/6G) and computational intelligence.

Machine Learning for Radio Resource Management and Optimization in 5G and Beyond

Edited by
Mariyam Ouaissa, Mariya Ouaissa,
Hanane Lamaazi, Khadija Slimani,
Ihtiram Raza Khan, and B. Sundaravadivazhagan

CRC Press
Taylor & Francis Group
Boca Raton London New York

CRC Press is an imprint of the
Taylor & Francis Group, an **informa** business

Designed cover image: ShutterStock Images

First edition published 2025
by CRC Press
2385 NW Executive Center Drive, Suite 320, Boca Raton FL 33431

and by CRC Press
4 Park Square, Milton Park, Abingdon, Oxon, OX14 4RN

CRC Press is an imprint of Taylor & Francis Group, LLC

ISBN: 9781032844732 (hbk)
ISBN: 9781032846552 (pbk)
ISBN: 9781003514336 (ebk)

DOI: 10.1201/9781003514336

Typeset in Minion
by codeMantra

Contents

Preface, vii

Editor biographies, xi

List of contributors, xiii

CHAPTER 1 ▪ Fundamentals of 5G and beyond networks 1

MARIYA OUAISSA AND MARIYAM OUAISSA

CHAPTER 2 ▪ Optimizing resource allocation in intelligent communication networks: Fundamentals and challenges 15

S. MAYUKHA AND R. VADIVEL

CHAPTER 3 ▪ Radio resource management for M2M communications in cellular networks 40

MARIYAM OUAISSA AND MARIYA OUAISSA

CHAPTER 4 ▪ Integrating blockchain for secure and efficient radio resource management in 5G and beyond networks 54

D. SUGANYA, S. SRI SAYE LAKSHMI, R. JAYALAKSHMI AND V. KAVITHA

CHAPTER 5 ▪ Federated learning for intelligent network management in 5G 73

B. SUNDARAVADIVAZHAGAN, N.A. NATRAJ, AND S. GOPINATH

CHAPTER 6 ■ Non-orthogonal multiple access wireless systems using deep learning 95

RUDRAKSH GOHIL AND S. DEEPA

CHAPTER 7 ■ Advancements in machine learning techniques for optimization of massive MIMO design 116

AVULA MAHATHI, C. KISHOR KUMAR REDDY, KARI LIPPERT AND MARLIA M. HANAFIAH

CHAPTER 8 ■ Predictive modeling of household power consumption using machine learning and meta-heuristic optimization technique 140

E. CHANDRA BLESSIE, B. SUNDARAVADIVAZHAGAN, V. KUMUTHA AND V. SUMESH

CHAPTER 9 ■ Intelligent reinforcement learning-based scheduling in 5G networks and beyond 156

ELARBI BADIDI, OMAR EL HARROUS AND HANANE LAMAAZI

CHAPTER 10 ■ AR/VR-based object detection for blind people using 5G communication 174

C. SUPRAJA AND T. KAVITHA

CHAPTER 11 ■ Exploring sentiment patterns in social media networks: The impact of AI, deep learning, and large models in the 5G landscape 198

OUSSAMA EL AZZOUZY, TARIK CHANYOUR, SAID JAI ANDALOUSSI, KHADIJA EL FELLAH, AND HABIBA BOUIJIJ

CHAPTER 12 ■ 5G and AI-based data fusion in intelligent networks 217

HANANE LAMAAZI AND ELARBI BADIDI

INDEX, 233

Preface

ADVANCES IN MOBILE COMMUNICATION systems have seen remarkable development in relation to the number of users as well as the data traffic. New generations of communication networks have introduced new techniques and continuously evolving architectures that can improve the quality of services and the performance of networks. The increasing density of traffic and the massive mobile connectivity from different devices and application will shape the next generations of mobile technologies.

Starting from 2020, the fifth-generation (5G) wireless communication networks are increasingly establishing themselves as the norm of communication despite the challenges they still face such as mass connectivity, ultra-reliability, and guaranteed low latency. 5G presents itself as another kind of mobile system model that supports high frequencies, extreme node densities, large-scale antennas, huge bandwidth, etc. Yet, to ensure the continuity and competitiveness in wireless communication systems, efforts from industry and academia have started to conceptualize the next generation of wireless communication systems, also known as the "sixth generation (6G)."

The sixth generation aims to provide revolutionary communication services and applications to meet the future demands of the 2030s by providing a global coverage, an enhanced spectral/energy/cost efficiency, a better intelligence level and security, etc.

Artificial intelligence/machine learning (AI/ML) approaches are promising tools to tackle the big challenges in wireless communications networks imposed by the increasing demands in terms of capacity, coverage, latency, efficiency, flexibility, compatibility, quality of experience, and silicon convergence. ML techniques can provide intelligent communication designs while addressing various problems ranging from signal detection, to channel modeling, network optimization, resource management, and multiple access.

This book offers an introduction to recent trends regarding 5G toward 6G communication networks. Moreover, it provides an overview of theoretical concepts and techniques of AI/ML used to meet the requirements and the challenges of radio resource management and optimization. This book presents comprehensive coverage of current and emerging waveform design, channel modeling, multiple access, random access, scheduling, and resource optimization for 5G wireless networks and beyond.

This book provides a comprehensive reference for researchers, scholars, and industry professionals in different fields related to mobile networks and intelligent systems. It can also be a hands-on resource for students interested in the fields of cellular networks (5G/6G) and computational intelligence.

Let's take a closer look at the specific themes and contributions of each chapter:

- The first and foundational chapter provides a comprehensive overview of the transition from 5G to 6G, detailing the fundamental principles, anticipated advancements, and the various challenges that need to be addressed to make 6G a reality.
- Chapter 2 explores the role of intelligent resource allocation techniques in enhancing network efficiency, reliability, and scalability.
- Chapter 3 reviews the random access procedure and the importance of the random access channel (RACH) and the challenges posed by congestion due to simultaneous access attempts.
- Chapter 4 presents an architectural framework for blockchain-based RRM and discusses deployment models (public, private, and consortium blockchains), and interoperability with existing systems.
- Chapter 5 examines federated learning's 5G network management challenges and prospects including its structure and kinds, including vertical, horizontal, and federated transfer learning.
- Chapter 6 centers on deep learning–NOMA integration with special attention given to areas like channel estimation, interference management, and dynamic resource allocation.
- Chapter 7 aims to discuss the progression of different machine learning techniques used for the optimization of the design of massive-MIMO design as well as the different limitations and challenges associated with their application.

- Chapter 8 comprehensively surveys the approach to predicting household power consumption using ML models enhanced with meta-heuristic optimization technique.
- Chapter 9 conducts the application of reinforcement learning (RL) for scheduling in 5G and beyond, focusing on its potential to solve resource allocation complexities. It covers basic RL strategies such as Markov decision processes (MDPs), Q-learning, and deep Q-networks (DQN), and details their application in developing scheduling solutions.
- Chapter 10 describes how smart vision glasses, NuEyes Pro, Envision, and Microsoft Hololens technologies can be used for disabled people and guide them using voice control with the help of ML and AI technology using 5G.
- Chapter 11 reviews recent advances in sentiment analysis in 5G-enabled social media networks, focusing on AI, deep learning, and large language models (LLMs) such as BERT and GPT.
- Chapter 12 showcases a comprehensive analysis of 5G and AI-based data fusion techniques in intelligent networks (INs). It briefly overviews the state-of-the-art features of these technologies, with a main focus on data fusion methods.

Editor biographies

Mariyam Ouaissa is currently an Assistant Professor in Networks and Systems at ENSA, Chouaib Doukkali University, El Jadida, Morocco. She earned her Ph.D. in 2019 from National Graduate School of Arts and Crafts, Meknes, Morocco, and her engineering degree in 2013 from the National School of Applied Sciences, Khouribga, Morocco. She is a communication and networking researcher and practitioner with industry and academic experience. Dr Ouaissa's research is multidisciplinary that focuses on the Internet of Things (IoT), M2M, WSN, vehicular communications and cellular networks, security networks, congestion overload problem, and the resource allocation management and access control.

Mariya Ouaissa is currently an Assistant Professor in Cybersecurity and Networks at the Faculty of Sciences Semlalia, Cadi Ayyad University, Marrakech, Morocco. She earned her Ph.D. in 2019 in Computer Science and Networks at the Laboratory of Modelisation of Mathematics and Computer Science from ENSAM-Moulay Ismail University, Meknes, Morocco. She is a Networks and Telecoms Engineer, graduated in 2013 from National School of Applied Sciences Khouribga, Morocco. She is a co-founder and IT consultant at IT Support and Consulting Center. She was working for the School of Technology of Meknes Morocco as a Visiting Professor from 2013 to 2021.

Hanane Lamaazi received an M.Sc. degree in Networks and Telecommunications from Chouaib Doukkali University and a Ph.D. degree in Network and Computer Sciences from Moulay Ismail University, Morocco, in 2013 and 2018, respectively. She was a postdoctoral fellow at the Center on Cyber-Physical Systems at Khalifa University, UAE, from 2019 to 2022. She is an Assistant Professor at the College of Information Technology at UAE University, UAE. Her research interests focus on the IoT, RPL routing protocol, edge computing, crowdsensing, and security.

Khadija Slimani earned her Ph.D. in Computer Science from the Faculty of Science at Ibn Tofail University in 2020. During her doctoral studies, she collaborated with the University of Technology of Belfort Montbéliard (UTBM), Montbéliard in France to conduct research with a focus on machine learning, deep learning, pattern recognition, and computer vision, specifically applied to academic emotion recognition. Upon the successful completion of her Ph.D. thesis, Dr. Slimani embarked on a postdoctoral journey at the University of Poitiers, assuming the role of a postdoctoral associate. In this capacity, her research revolved on objects DRI (Detection, Recognition, and Identification), while employing machine learning and deep learning methodologies to enhance video content filtering for optimal security. Her contributions extended to diverse engineering schools in Paris, where she undertook teaching responsibilities across modules spanning data science, deep learning, machine learning, databases, and computer vision. She is currently an Associate Professor at the Graduate School of Automatic Electronic Computing in Paris, France.

Ihtiram Raza Khan is working as senior academician at Jamia Hamdard, Delhi, He has more than 26 years of experience and earned his Ph.D. in the field of software engineering and neural networks. His research interests include software engineering, computer graphics, machine and deep learning, big data, analytics, cybersecurity, and IoT. He has been actively involved in training and placement activities as Head and has offered consultancies to more than 15 companies. He has more than 20 international and Indian patents and copyrights against his name. He has written more than 20 books and 30 book chapters, and more than 75 research papers in SCI/Scopus/Springer and peer-reviewed journals.

B. Sundaravadivazhagan is an experienced researcher and educator in the field of information and communication engineering. He has more than 21 years of experience in teaching and research and earned his Ph.D. in Information and Communication Engineering from Anna University in Chennai in 2016. He is a member of various professional bodies such as IEEE, ISACA, ISTE, and ACM, and has published more than 40 research articles in SCI and Scopus journals. He has also served as a resource person, keynote speaker, and advisory committee member in more than 20 international and national conferences. He has received two research grants from the Ministry of Higher Education, Research and Innovation, The Research Council (TRC), Oman. His research interests include IoT, AI, machine learning, deep learning, cloud computing, networks and security, wireless networks, and MANET.

Contributors

Said Jai Andaloussi
Hassan II University, Morocco

Elarbi Badidi
UAE University, Al Ain, United Arab Emirates

E. Chandra Blessie
Coimbatore Institute of Technology, India

Habiba Bouijij
Mohammed V, Morrocco

Tarik Chanyour
Hassan II University, Morocco

S. Deepa
Christ University, India

Oussama El Azzouzy
Hassan II University, Morocco

Khadija El Fellah
Ibn Tofail University, Morocco

Omar El Harrous
UAE University, United Arab Emirates

Rudraksh Gohil
Christ University Bangalore, India

S. Gopinath
Karpagam Institute of Technology, India

Marlia M. Hanafiah
University of Kebangsaan Malaysia, Malaysia

R. Jayalakshmi
Rajiv Gandhi college of Engineering and Technology, India

T. Kavitha
Vel Tech Rangarajan Dr. Sagunthala R&D Institute of Science and Technology, India

V. Kavitha
Rajiv Gandhi college of Engineering and Technology, India

V. Kumutha
Avinashilingam Institute of Home Sciences for Women, Coimbatore, India

Hanane Lamaazi
UAE University, United Arab Emirates

Kari Lippert
University of South Alabama, USA

Avula Mahathi
Stanley College of Engineering and Technology for Women, India

S. Mayukha
Bharathiar University, India

N.A. Natraj
Symbiosis International (Deemed University), India

Mariya Ouaissa
Cadi Ayyad University, Morocco

Mariyam Ouaissa
Chouaib Doukkali University, Morocco

C. Kishor Kumar Reddy
Stanley College of Engineering and Technology for Women, India

S. Sri Saye Lakshmi
Rajiv Gandhi college of Engineering and Technology, India

D. Suganya
Rajiv Gandhi college of Engineering and Technology, India

V. Sumesh
Coimbatore Institute of Technology, India

B. Sundaravadivazhagan
University Technology and Applied Sciences-Al Mussana, Oman

C. Supraja
Vel Tech Rangarajan Dr. Sagunthala R&D Institute of Science and Technology, India

R. Vadivel
Bharathiar University, India

CHAPTER 1

Fundamentals of 5G and beyond networks

Mariya Ouaissa and Mariyam Ouaissa

1 INTRODUCTION

Communication is the very foundation of all human civilizations throughout history. The modern era introduced wireless telecommunications, enabling the emergence of contemporary society. However, as technological prowess evolves exponentially, so do the needs and demands of consumers. Nowadays, several application areas such as the Internet of Things (IoT), artificial intelligence (AI), and data mining require a fast, reliable, efficient, and wide-bandwidth Internet connection. In many cases, telecommunications devices are located in places where direct access to a local network is unavailable. Fortunately, mobile networks cover vast territories, allowing these devices to remain connected to the world at all times [1].

The first generation of mobile telephony (1G) featured analog operation and bulky communication terminals, vastly different from today's devices. In the 1980s, the National Center for Telecommunications Studies conducted significant research, laying the groundwork for the Global System for Mobile Communication (GSM) standard. This standard facilitated the development of mobile telephony, particularly in Europe, using the 900 MHz and 1,800 MHz bands, and different frequencies in the United States. From the second generation (2G) onward, the technology became entirely digital, allowing phones to exchange Short Message Service (SMS) text messages [2]. A version (2.5G) with faster speeds, around 40 kbit/s, using

DOI: 10.1201/9781003514336-1

the General Packet Radio System (GPRS) standard, emerged. Likewise, the Enhanced Data Rates for Global Evolution (EDGE) version, with speeds up to 384 kbit/s, was also introduced.

The advent of 3G, a decade after 2G, revolutionized mobile phone usage, offering speeds from 384 kbit/s to 2 Mbit/s, and later up to 42 Mbit/s. Multimedia messages such as Multimedia Messaging Service (MMS), emails, photos, and videos could be exchanged at speeds of around 1.9 Mbit/s. Following the International Mobile Telecommunications for the year 2000 (IMT-2000) specifications by the International Telecommunications Union (ITU), several advantages were highlighted, including variable speeds depending on coverage (fixed, moving, etc.) and universal compatibility to allow Internet access. From 2005, the Universal Mobile Telecommunications System (UMTS) standard transformed terminals into web servers, providing multimedia services like television and videoconferencing. By early 2008, nearly 300 million mobile terminals were in service. Variants like 3G+ (10 Mbit/s) and 3G++ (42 Mbit/s) also emerged [3]. 4G uses the more efficient long-term evolution (LTE) standard, offering speeds of 150 Mbit/s. 4G+ uses the LTE advanced standard, providing theoretical speeds of 1 Gbit/s, often closer to 300 Mbit/s. 4G has contributed to the development of mobile Internet, video usage, and videoconferencing. 5G is characterized by speeds up to ten times higher than 4G (>1 Gbit/s), with latency (response time) reduced by 90% and connection density increased tenfold. 5G connects people, objects, data, applications, networks, transport systems, and territories [4].

While 5G is beginning to be deployed across all continents, research on 6G is already underway in relevant laboratories and companies. The first 6G networks are expected to be operational by 2028, aiming to reach speeds of 1,000 Gbit/s in the terahertz frequency range. 6G will be built with new technologies and capabilities to meet future communication needs [5].

This chapter aims to thoroughly explore the shift from 5G to 6G, outlining the core principles, expected advancements, and the various challenges that must be overcome to realize 6G. The structure of this chapter is organized as follows: Section 2 provides an overview of 5G system. Section 3 presents the evolution from 5G to 6G. In Section 4, we describe the challenges and advantages of 6G. Conclusions are drawn in Section 5.

2 OVERVIEW OF 5G

5G is the new generation of mobile telephony standards, set to connect every physical object in the world to the Internet under the new concept of the IoT. Absolutely everything, from autonomous vehicles to surgical robots, augmented reality devices, drones, and mobile communications in general, will utilize a new bandwidth range between 6 GHz and 300 GHz [6]. Part of this spectrum will be licensed, but a large portion will remain unlicensed. Unlike an evolution of previous networks, 5G represents a new technological foundation developed to meet growing customer demands. For the first time, the physical infrastructure will be separated from the logical or "virtual" infrastructure. A Software Defined Network (SDN) will manage the Internet network, making it programmable to support new applications and Internet services [7].

2.1 Principle

5G uses ultrahigh broadcast frequencies (UHF) between 6 and 300 GHz, specifically millimeter waves, to quickly transfer huge amounts of data. However, the broadcast range is limited, which is addressed by using beamforming technology through multiuser multiple input, multiple output (MU-MIMO). This technology allows the continuous broadcasting of data to multiple devices and ensures gigabit-level connections even during multicell handover processes [8].

2.1.1 Millimeter waves

Millimeter waves are radio waves covering frequencies from 30 to 300 GHz.

2.1.2 Microcells

Microcells increase spectral efficiency in an area and reduce cell size. With fewer users per cell, each user gets more spectrum. Reducing cell sizes and reusing spectrum significantly increases the total network capacity.

2.1.3 Massive MIMO

The need for high speeds in next-generation mobile communications has led to the use of MIMO technology, which increases the number of transmitting and receiving antennas. To reduce signal fading caused by multipath propagation and channel noise, the diversity technique can be used. This involves placing one transmitting antenna and multiple receiving antennas. However, integrating multiple antennas into the limited volume of mobile devices is a challenge. Strong isolation between the antennas is necessary to maximize diversity gain and increase the overall transmission capacity of the system.

2.1.4 Beamforming

Beamforming allows the adaptation of the antenna array's radiation pattern according to a specific model.

2.1.5 Full-duplex

Full-duplex describes a serial or network connection that allows simultaneous reception and transmission of data.

2.2 5G network architecture and its services

A 5G network comprises a 5G Access Network (AN) and a 5G core network. The AN includes the next-generation radio access network (NG-RAN), utilizing the new 5G New Radio (NR) interface, along with potentially non-3rd Generation Partnership Project (3GPP) access networks linked to the 5G core network. These network components are interconnected via an underlying TCP/IP transport network that facilitates quality of service (QoS) functionalities [9].

As illustrated in Figure 1.1, the 5G network establishes connectivity between user equipment (UE) and external data networks. The PDU session embodies the 5G connectivity service, comprising a series of NG tunnels within the 5G core network (5GC) alongside one or more radio bearers on the radio interface. This collection of pathways links the UE to its control functions and facilitates exchange of user traffic with external data networks (Figure 1.2).

A protocol data unit (PDU) session resembles an LTE EPS bearer but differs in its QoS model and supported user data units. Beyond IP packets, a PDU session can carry Ethernet frames or other non-IP packets, thereby enabling Layer-2 communication among UE groups. The 5G framework

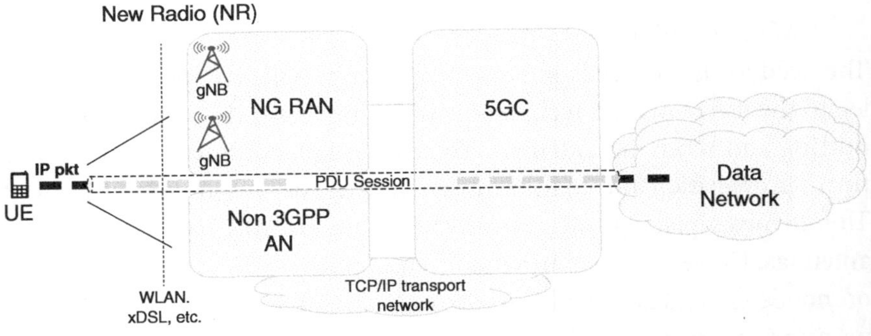

FIGURE 1.1 5G network architecture.

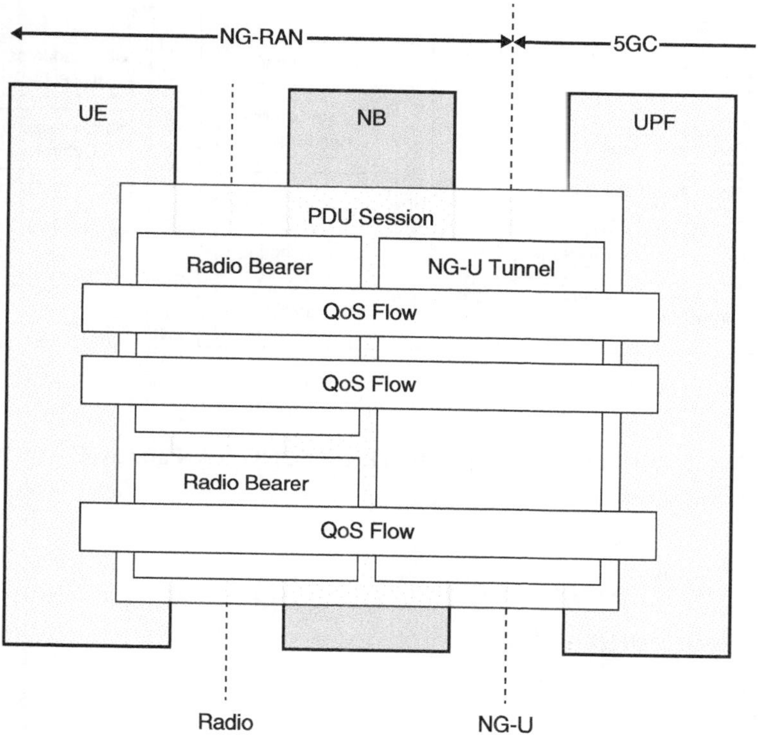

FIGURE 1.2 PDU session and QoS flow.

introduces the concept of QoS flow, where each flow represents the finest level of QoS differentiation within a PDU session.

Figure 1.3 delineates the roles within the 5G network, distinguishing between functions performed in the NG-RAN and 5G core. The NG-RAN operates at the network edge, responsible for initiating, maintaining, and terminating segments of PDU sessions across the radio interface. It tackles challenges like physical constraints (e.g., interference, energy efficiency), inter-gNB handovers, and session multiplexing (scheduling). Meanwhile, the 5GC oversees the remainder of the PDU sessions and manages non-radio access processes such as mobility management, IP address allocation, and security [10].

1.2.2.1 New generation radio access network

Figure 1.4 illustrates the NG-RAN, comprising multiple 5G base stations known as gNBs, interconnected with the 5GC through logical interfaces. These gNBs can establish connections via the Xn interface to enhance

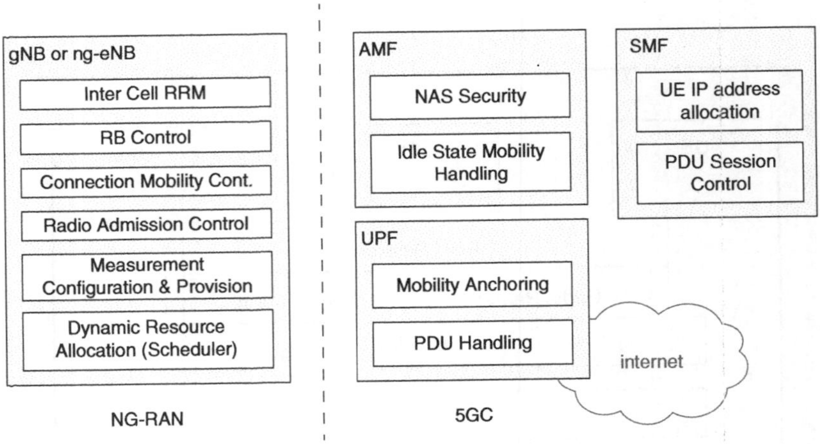

FIGURE 1.3 The functional separation between NG-RAN and 5GC.

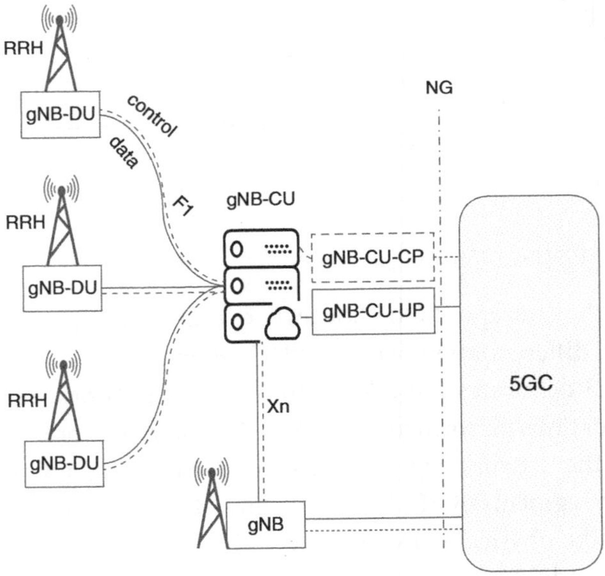

FIGURE 1.4 Architecture of NG-RAN.

mobility, including handover capabilities and manage intercellular interference. The gNB architecture is sometimes decentralized, featuring a central unit (gNB-CU) that coordinates with one or more distributed unit (gNB-DU) and connects to a remote radio head (RRH). The gNB-CU is subdivided into control plane functions (gNB-CU-CP) and user plane functions (UPFs) (gNB-CU-UP) to efficiently manage network operations [11].

2.2.2 5G core network

The evolution of network node functions from previous generations has shaped a 5G architecture defined by network functions exposed as services. Similar to the NG-RAN, the 5G architecture segregates into control and user planes. In the user plane, one or more UPFs manage packet transfer across NG-U tunnels forming the PDU session. All other network functions reside in the control plane.

A significant departure from past architectures is the interface modeling transition from "bit-oriented point-to-point" to "web-oriented service-based" approaches. The 5GC architecture leverages service-based definitions of procedures to facilitate reusability. Previous generations relied on standardized point-to-point interfaces, real or logical, between 2G, 3G, and 4G network entities using specialized bit-oriented protocols. In contrast, interactions among control plane entities in 5G core utilize service-based interfaces supported by web technologies such as HTTP/2, REST, and JSON [12] (Figure 1.5).

The network comprises various nodes, each serving specific functions:

1. User Plane (UP): Manages NG-U tunnel transmission and related data path services, including anchoring transfers, QoS enforcement, and traffic policies.

2. Session Management Function (SMF): Controls PDU sessions by configuring NG tunnels, assigning IP addresses via DHCP, and

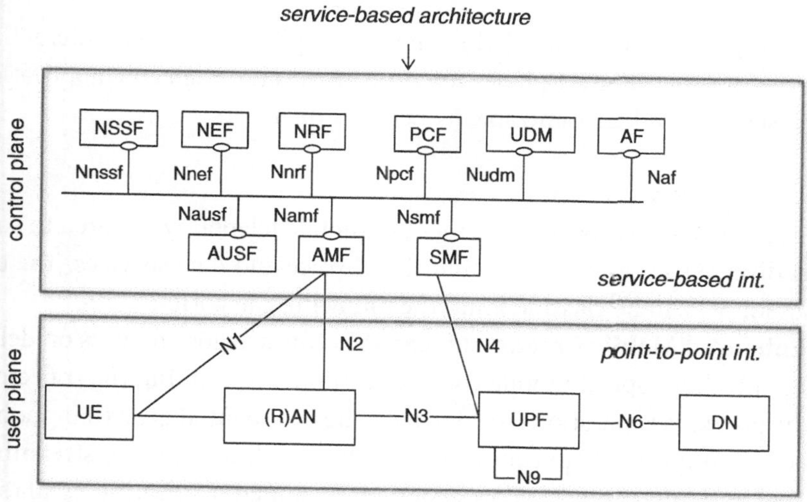

FIGURE 1.5 5G system architecture (no-roaming).

directing traffic to third party or edge clouds. Multiple SMFs can be associated with an UE, limited to one per slice.

3. Access and Mobility Management Function (AMF): Handles all 5GC signaling to and from the UE, facilitating user access and managing mobility across multiple slices.
4. Authentication Server Function (AUSF): Provides authentication services for both 3GPP and non-3GPP accesses.
5. Unified Data Management (UDM): Acts as a repository for information such as identifiers, AMF details, and SMF assignments for the current session.
6. Policy Control Function (PCF): Manages unified rule sets (QoS, filtering, load management, etc.) for other control plane functions, including SMF.
7. Network Slice Selection Function (NSSF): Selects network slices that serve the UE's requirements and identifies the optimal AMF for connectivity.
8. Network Exposure Function (NEF): Exposes network capabilities and UE-related events to third-party application functions and other entities.
9. Network Repository Function (NRF): Discovers network function instances in response to NF discovery requests from NF instances.
10. Application Function (AF): Supports application servers interacting with other control plane functions, tailored for specific application services and customizable by network operators.

2.2.3 5G services

5G represents a pivotal advancement in mobile telephony, tailored to support diverse use cases beyond traditional voice and data services. The use cases for 5G can be classified into the following categories:

Enhanced Mobile Broadband (eMBB): This category focuses on delivering ultrahigh-speed mobile communications, continuing the trajectory set by previous generations to cope with the exponential growth in mobile data consumption. Applications include high-definition video streaming, virtual and augmented reality experiences, aimed at meeting escalating demands for data volume and transfer speeds.

Massive Machine Type Communications (mMTC): mMTC revolves around facilitating communication between a vast array of objects, forming the backbone of the IoT. This category is designed to manage a large volume of connections, potentially reaching up to 1 million connections per square kilometer. Each connection typically involves smaller data volumes and requires less stringent transfer speed requirements. Applications range from sensor networks in smart cities to support various services and functionalities.

Ultra-Reliable Low Latency-Communications (URLLC): URLLC addresses critical communication needs that demand extremely high reliability and minimal latency. Key applications include autonomous vehicles, as well as communication systems for security and emergency services. These applications necessitate uninterrupted communication capabilities and ultrafast transmission speeds to ensure safety and efficiency in real-time scenarios.

2.3 5G standardization

The specification of each new generation of mobile telephony is governed by two main bodies: the ITU and the 3GPP. The ITU, a United Nations agency responsible for information and communication technologies, conducts studies through its "Working Party 5D," which addresses technical issues and manages radio communications [13].

Since 2013, this group has been defining the new IMT standard, IMT-2020 (5G), similar to how the 4G (IMT-Advanced) standard was defined in the early 2000s. Concurrently, the 3GPP, established in 1998, collaborates with seven standardization organizations, industry players, associations, and public entities to develop and maintain technical specifications for mobile telephony standards. When the ITU defines a new standard, the 3GPP concurrently works on technical solutions to meet the ITU's objectives.

The 3GPP released the first standard for 5G, known as Release 15, in December 2016, initiating the definition of a new architecture starting from March 2017, including a new air interface called NR. The first version of the standard was validated in September 2018, followed by the publication of the second release, 3GPP Release 16, in March 2020. The transition from one generation to the next occurs progressively, with LTE continuing to evolve alongside NR. Both standards are expected to be highly complementary. The objectives set for 5G have been incorporated into 3GPP Releases 13, 14, and 15, which not only cater to 5G but also encompass advancements in 4G (referred to as 4.9G or LTE Advanced Pro).

The 3GPP standards group, with contributions from global participants, is responsible for defining the 5G wireless standard. Version 15 (Release 15) of the 3GPP standard, introduced in September 2018, marked the initial standardization of 5G. The standards body has reached agreements on several key aspects:

- Multiple waveform types: Including cyclic prefix OFDM and SC-FDMA (single carrier frequency division multiple access) (used in LTE), with filtered OFDM and windowed OFDM under study as additional spectrum options.
- Two types of propagation channels: Simulated across frequencies from 500 MHz to 100 GHz.
- Variable spacing between carriers: Resulting in variable lengths of OFDM symbols.
- Multiple coding schemes: Incorporating low density parity check (LDPC) coding for data channels and polar coding for control channels.

3 5G TECHNOLOGY AND THE 6G ROADMAP

The benefits of 5G extend beyond increased data throughput, encompassing reduced battery consumption, lower outage probability, enhanced security, and improved coverage. Data rates at the cell edge reach up to 1 Gbit/s in mobility, and the technology is designed to be nonharmful to human health. Deployment costs are expected to decrease due to the implementation of smart beam antenna systems, along with advanced data coding and modulation techniques that enable comprehensive multimedia capabilities surpassing those of 4G [14].

To achieve these goals, 5G utilizes new frequency bands in addition to those used by previous generations. Furthermore, in the era of the IoT, pervasive connectivity facilitates the networking of everyday objects. This integration involves embedding each object in a network, creating a highly distributed system of interconnected devices.

Moreover, the Internet of Everything (IoE) integrates people, processes, data, and things to enhance network connections, transforming information into actionable insights that create new capabilities and richer experiences for individuals and businesses. In simpler terms, IoE refers to the intelligent connectivity of people, processes, data, and objects, where billions of sensor-equipped objects communicate over public or private networks using standard or proprietary protocols [15].

These new IoT and IoE paradigms necessitate optimal data transmission methods, emphasizing system integration to manage vast amounts of data while maintaining high transfer quality. Furthermore, to sustain the competitive edge of wireless networks, industrial and academic collaborations are already conceptualizing the next generation of wireless communication systems (6G) to meet the communication demands of the 2030s [16].

Several 6G communication scenarios are emerging:

- **Enhanced Mobile Broadband (eMBB-Plus):** A successor to 5G eMBB, focusing on delivering high-quality user experiences (QoE) in data usage.
- **Big Data Computing and Communication (BigCom):** Aims to provide extensive coverage in urban and remote areas, balancing resource usage to enable high-data-rate communication between subscribers.
- **Secure Ultra-Reliable Low-Latency Communications (SURLLC):** Provides ultra-reliable, secure communications with latency under 0.1 ms, crucial for applications like vehicular communications in 6G.
- **Three-Dimensional Integrated Communications (3D-InteCom):** Incorporates communication nodes in the third dimension, including satellites, unmanned aerial vehicles, and underwater communications.
- **Unconventional Data Communications**: Includes technologies like holographic, tactile, and human communications, which are anticipated in 6G without detailed specification at this stage.

4 ADVANTAGES AND CHALLENGES OF 6G

The primary objective of advancing to a new technology or preparing for the next generation is to achieve benefits that previous iterations cannot provide. While the exact performance of 6G remains uncertain until its development is further advanced, many experts anticipate specific outcomes for this next evolution in wireless technology [17]. Overall, 6G is expected to offer six key advantages over 5G:

- Increased capacity through the inclusion of sub-THz spectrum, spanning from 30 to 300 GHz.
- Improved coverage and reliability with spectrum expansion, potentially reaching up to 99.99999% availability (7 9's).
- Peak data rates increased to 1 terabit per second.
- Reduced latency aiming for microsecond-level latency.
- Enhanced spectral efficiency, potentially exceeding 100 bps/Hz.
- Increased mobility support up to speeds of 1,000 km/h (600 mph).

Each generation of cellular communications technology represents a significant leap forward in research and development, driving innovation and expanding use cases. 6G is no exception [18]. However, some of the foundational technologies required for 6G do not exist today. Researchers are actively exploring new approaches to address the challenges posed by 6G:

- Higher communication frequencies, potentially above 100 GHz and even reaching toward 1 THz. This will necessitate new radio frequency (RF) designs and possibly new semiconductor materials.
- Faster processing capabilities to handle speeds exceeding 1 terabit per second (Tbps). This may involve optical processing techniques or advancements in quantum computing.
- Energy consumption management and reduction. While 5G networks are 90% more energy-efficient than their 4G counterparts based on data bits per kilowatt, the substantial increase in network density and traffic could offset these gains. Consequently, 5G networks may end up consuming four to five times more net energy than 4G. Given that 6G networks are expected to be denser and faster than 5G, managing power consumption will be a significant challenge.

5 CONCLUSION

The evolution from 1G to 5G networks has marked a transformative leap in telecommunications, enabling unprecedented speeds, reliability, and connectivity. The fundamental technologies of 5G, such as millimeter wave spectrum, massive MIMO, and network slicing, have set new benchmark for mobile broadband, IoT connectivity, and URLLC. As we look toward the future, the emergence of 6G promises even greater advancements, including higher capacities, reduced latencies, and enhanced energy efficiencies, paving the way for innovations that will redefine industries and daily life. The journey from 5G to 6G exemplifies the continuous pursuit of pushing technological boundaries to meet the evolving demands of a connected world, ensuring that networks not only support but also catalyze future innovations across diverse sectors.

REFERENCES

1. C. Sudhamani, M. Roslee, J. J. Tiang, and A. U. Rehman, "A survey on 5G coverage improvement techniques: Issues and future challenges," *Sensors*, vol. 23, no. 4, 2023, p. 2356.
2. M. Ouaissa, A. Rhattoy, and I. Chana, "New security level of authentication and key agreement protocol for the IoT on LTE mobile networks," In *2018 6th International Conference on Wireless Networks and Mobile Communications (WINCOM)*, Marrakesh, Morocco, pp. 1–6, IEEE, 2018.
3. M. Ouaissa, M. Benmoussa, A. Rhattoy, M. Lahmer, and I. Chana, "Performance analysis of random access mechanisms for machine type communications in LTE networks," In *2016 International Conference on Advanced Communication Systems and Information Security (ACOSIS)*, Marrakesh, Morocco, pp. 1–6, IEEE, 2016.
4. M. Banafaa, Ö. Pepeoğlu, I. Shayea, A. Alhammadi, Z. Shamsan, M. A. Razaz, M. Alsagabi, and S. Al-Sowayan, "A comprehensive survey on 5G-and-beyond networks with UAVs: Applications, emerging technologies, regulatory aspects, research trends and challenges," *IEEE Access*, vol. 12, 2024, pp. 7786–7826.
5. A. Dogra, R. K. Jha, and S. Jain, "A survey on beyond 5G network with the advent of 6G: Architecture and emerging technologies," *IEEE Access*, vol. 9, 2020, pp. 67512–67547.
6. A. Sufyan, K. B. Khan, O. A. Khashan, T. Mir, and U. Mir, "From 5G to beyond 5G: A comprehensive survey of wireless network evolution, challenges, and promising technologies," *Electronics*, vol. 12, no. 10, 2023, p. 2200.
7. M. Wu, F. R. Yu, and P. X. Liu, "Intelligence networking for autonomous driving in beyond 5G networks with multi-access edge computing," *IEEE Transactions on Vehicular Technology*, vol. 71, no. 6, 2022, pp. 5853–5866.
8. M. Houmer, M. Ouaissa, and M. Ouaissa, "Secure authentication scheme for 5G-based V2X communications," *Procedia Computer Science*, vol. 198, 2022, pp. 276–281.
9. M. Ouaissa and M Ouaissa, "An improved privacy authentication protocol for 5G mobile networks," In *2020 International Conference on Advances in Computing, Communication & Materials (ICACCM)*, Dehradun, India, pp. 136–143, IEEE, 2020.

10. G. Thirunavukkarasu and G. Murugesan, "A comprehensive survey on air-interfaces for 5G and beyond," In *2019 10th International Conference on Computing, Communication and Networking Technologies (ICCCNT)*, Kanpur, India, pp. 1–7, IEEE, 2019.
11. M. Ouaissa, M. Houmer, and M. Ouaissa, "An enhanced authentication protocol based group for vehicular communications over 5G networks," In *2020 3rd International Conference on Advanced Communication Technologies and Networking (CommNet)*, Marrakech, Morocco, pp. 1–8, IEEE, 2020.
12. N. Trabelsi, L. C. Fourati, and C. S. Chen, "Interference management in 5G and beyond networks: A comprehensive survey," *Computer Networks*, vol. 239, 2023, p. 110159.
13. M. Houmer, M. Ouaissa, and M. Ouaissa, "Secure authentication scheme for 5G-based V2X communications," *Procedia Computer Science*, vol. 198, 2022, pp. 276–281.
14. S. Rajoria and K. Mishra, "A brief survey on 6G communications," *Wireless Networks*, vol. 28, no. 7, 2022, pp. 2901–2911.
15. C. De Alwis, A. Kalla, Q. V. Pham, P. Kumar, K. Dev, W. J. Hwang, and M. Liyanage, "Survey on 6G frontiers: Trends, applications, requirements, technologies and future research," *IEEE Open Journal of the Communications Society*, vol. 2, 2021, pp. 836–886.
16. A. T. Jawad, R. Maaloul, and L. Chaari, "A comprehensive survey on 6G and beyond: Enabling technologies, opportunities of machine learning and challenges," *Computer Networks*, vol. 237, 2023, p. 110085.
17. Y. Lu and X. Zheng, "6G: A survey on technologies, scenarios, challenges, and the related issues," *Journal of Industrial Information Integration*, vol. 19, 2020, p. 100158.
18. S. Aneesh and A. N. Shaikh, "A survey for 6G network: Requirements, technologies and research areas," In *2023 2nd International Conference on Edge Computing and Applications (ICECAA)*, Namakkal, India, pp. 166–171, IEEE, 2023.

CHAPTER 2

Optimizing resource allocation in intelligent communication networks

Fundamentals and challenges

S. Mayukha and R. Vadivel

1 INTRODUCTION

Cloud computing is the combination of large-scale parallel [1], distributed, and multi-tenant models which involves several technologies such as virtualization and automatic computing that furnish a pool of resources to various cloud consumers [2]. Cloud manages the data center which is a collection of servers where application is placed and is retrieved via the internet. Cloud data centers offer unlimited resources provision to their applications based on on-demand self-service. However, cloud maintains 1,000–10,000 of data servers in data center; the energy consumption of data centers have increased to $140 billion until 2020, and it increases every year that leads to high consumption of power, large amount of energy, and high operation cost for user-requested services worldwide [3, 4]. The resource allocation in cloud data center invokes several challenges

DOI: 10.1201/9781003514336-2

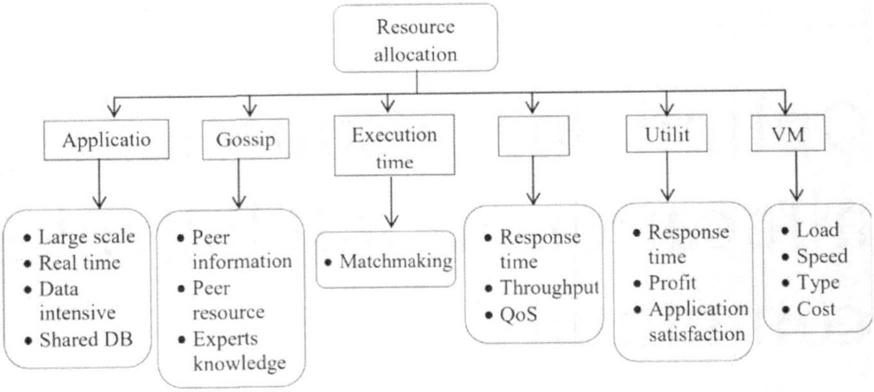

FIGURE 2.1 **Resource allocation based on different parameters.**

such as energy efficiency, virtual machine (VM) migration and server consolidation. With the ever-growing increase in energy consumption of data center, energy management has great consideration in cloud research that can be concentrated in cooling systems and servers. Figure 2.1 explains the classification of resource allocation strategies based on several servers, applications and infrastructure.

In data centers, overutilization of resources is one of the major reasons of power inefficiency. To reduce the peak demand, that is, overutilization, resources are underutilized in most of the time. The average utilization of servers ranges from 10% to 30% for huge data centers, which results in a situation where considerable capacity of data center is idle. Hence, VM consolidation, placement, and migration techniques have been effectively applied to improve the server power efficiency for the servers which are not energy proportional. An effective resource allocation process satisfies the following constraints [5]:

Location Independence: Data centers have been established separately around the world. Energy cost and performance cost depend on the location of data center.

Dynamic Resource Scalability: To maintain the greater number of users on cloud, dynamic resource can scale up and down the bounds of virtualized resources.

Unnecessary Wastage Elimination: Terminating the idle physical machines (PMs) reduces power consumption and resource redundancy.

Efficient Resource Utilization: An effective scheme must finely allocate resource for high utilization.

1.1 Benefits of resource allocation

- In resource allocation, users do need not to install either software or hardware to access the application over internet.
- Cloud providers share their resources over internet during resource scarcity. Virtualization environments allow VMs to share the physical resources that include memory, disk, network devices, and so on.
- Resource allocation systems avoid server overloaded/underloaded problem while predicting the number of servers effectively.
- Resource allocation supports easy access to the entire resources and manages the workloads of different resources.
- Resource allocation provides several benefits such as reducing energy consumption as well as time consumption, minimizing overall risk and decreasing cost.

2 DYNAMIC RESOURCE MANAGEMENT

Dynamic resource management (DRM) refers to the process of dynamically allocating and reallocating resources in a network or system based on real-time conditions and demands. This concept is crucial in various fields, including telecommunications, computer networks, cloud computing, and distributed systems. Intelligent networks use algorithms and protocols to dynamically allocate resources based on real-time demand and network conditions. This includes allocating bandwidth, spectrum, and computational resources efficiently. The primary goal of DRM is to optimize resource utilization, improve performance, and enhance the overall efficiency of the system [6].

2.1 Benefits of dynamic resource management

- **Improved efficiency**: By dynamically reallocating resources based on real-time needs, DRM reduces waste and improves overall system efficiency.
- **Enhanced scalability**: Systems using DRM can scale resources seamlessly to accommodate growing demands or sudden spikes in workload.

- **Better quality of service (QoS)**: DRM helps maintain consistent performance levels and meets service-level agreements (SLAs) by prioritizing resource allocation for critical applications.
- **Cost optimization**: Efficient resource management reduces unnecessary expenditures on over-provisioned resources, optimizing operational costs.

2.2 Key aspects of dynamic resource management

- **Real-time adaptation**: DRM involves continuously monitoring system parameters such as traffic load, bandwidth utilization, latency, and other performance metrics. Based on these real-time data, resources are allocated or adjusted to meet current demands effectively.
- **Flexibility and scalability**: DRM allows systems to scale resources up or down based on fluctuating demands. This scalability ensures that resources are efficiently utilized without over-provisioning or under-provisioning.
- **Automation and intelligence**: DRM often integrates automation and intelligent algorithms (such as machine learning (ML) or artificial Intelligence (AI) to make informed decisions about resource allocation. These algorithms can predict future demands, optimize resource distribution, and improve overall system responsiveness.
- **Quality of service**: DRM plays a crucial role in ensuring QoS by prioritizing resource allocation for critical applications or services. This involves setting policies and rules that govern how resources are allocated based on SLAs or user requirements.
- **Resource virtualization**: In virtualized environments (e.g., cloud computing, virtual networks), DRM manages virtual resources dynamically across physical infrastructure. This allows for efficient resource sharing and allocation among multiple users or applications.
- **Energy efficiency**: DRM can contribute to energy savings by dynamically adjusting resource usage patterns to minimize power consumption during low-demand periods or by consolidating workloads onto fewer servers.

2.3 Applications of dynamic resource management

- **Telecommunications**: In mobile networks, DRM optimizes spectrum allocation, adjusts bandwidth dynamically, and manages network resources to handle varying traffic loads.
- **Cloud computing**: DRM in cloud environments dynamically allocates VMs, storage, and network resources to optimize performance and cost-efficiency based on workload demands.
- **Distributed systems**: DRM ensures efficient utilization of computing resources in distributed systems by dynamically balancing tasks across nodes and adjusting resource allocations as workload patterns change.
- **Internet of Things (IoT)**: In IoT networks, DRM manages resources across a vast number of connected devices, ensuring optimal data processing, communication bandwidth, and energy usage.

2.4 Artificial intelligence

AI refers to the simulation of human intelligence in machines that are programmed to think, learn, and problem-solve like humans. AI encompasses a broad range of techniques and approaches aimed at enabling computers and systems to perform tasks that traditionally require human intelligence (Figure 2.2). AI plays a pivotal role in optimizing resource allocation in intelligent communication networks. These networks include modern telecommunications infrastructure, such as 5G and beyond, and the IoT ecosystems. AI's role in optimizing resource allocation in intelligent communication networks is transformative, thereby offering numerous benefits in terms of efficiency, performance, and cost savings. By leveraging advanced ML and predictive analytics, AI can dynamically manage resources in real time, adapt to changing conditions, and enhance the overall user experience. However, successful implementation requires addressing challenges related to complexity, data quality, security, and compliance [7].

- **Machine learning**
  - ML is a subset of AI focused on developing algorithms that allow machines to learn from data and make predictions or decisions based on that learning. It includes techniques like supervised learning, unsupervised learning, and reinforcement learning (RL).

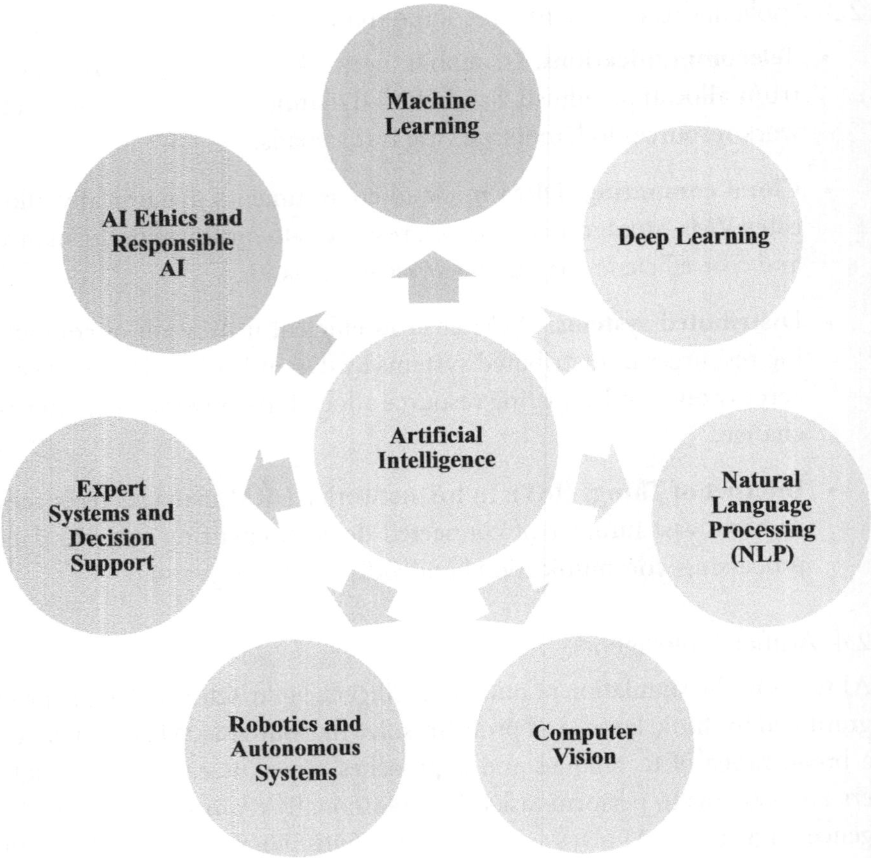

FIGURE 2.2 Artificial intelligence.

- **Deep learning**
    - Deep learning is a specialized form of ML that uses neural networks with many layers (deep neural networks) to analyze data. It has been particularly successful in tasks such as image and speech recognition, natural language processing (NLP), and autonomous driving.

- **Natural language processing**
    - NLP focuses on enabling computers to understand, interpret, and generate human language in a way that is both meaningful and contextually relevant. Applications include language translation, sentiment analysis, and chatbots.

- **Computer vision**
  - Computer vision involves teaching computers to interpret and understand the visual world through digital images or videos. It is used in applications such as object detection, facial recognition, and medical image analysis.
- **Robotics and autonomous systems**
  - AI plays a crucial role in robotics by enabling robots and autonomous systems to perceive their environment, make decisions, and perform tasks autonomously. This includes applications in manufacturing, healthcare, agriculture, and space exploration.
- **Expert systems and decision support**
  - Expert systems use AI techniques to mimic the decision-making ability of a human expert in a specific domain. They are used in fields such as healthcare diagnostics, financial planning, and troubleshooting complex systems.
- **AI ethics and responsible AI**
  - With the increasing deployment of AI technologies, ethical considerations around fairness, transparency, accountability, and bias mitigation have become crucial. Responsible AI frameworks aim to ensure that AI systems operate ethically and contribute positively to society [8].

2.5 Benefits of AI in optimizing resource allocation

- **Improved efficiency:** AI algorithms can optimize the allocation of resources in real time, leading to better utilization and reduced waste.
- **Enhanced performance:** By dynamically adjusting network parameters, AI helps maintain high performance and service quality.
- **Scalability:** AI systems can handle the complexity of large-scale networks, making them suitable for modern, highly interconnected environments.
- **Cost reduction:** Optimized resource allocation reduces operational costs, including energy consumption and maintenance expenses.
- **Adaptability:** AI systems can quickly adapt to changing network conditions and user demands, thereby ensuring continuous optimization [9].

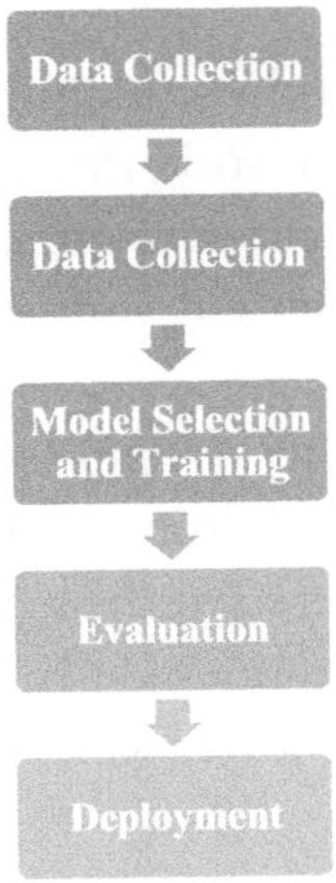

FIGURE 2.3 Machine learning workflow.

2.6 Machine learning

It is a branch of AI that focuses on enabling systems to learn and improve from experience automatically without being explicitly programmed. It involves developing algorithms and models that allow computers to learn patterns and make decisions based on data (Figure 2.3).

- **Data collection**: Gathering relevant data from various sources.
- **Data preprocessing**: Cleaning, transforming, and preparing data for analysis.
- **Model selection and training**: Choosing an appropriate ML algorithm, training the model on the data, and optimizing its performance.
- **Evaluation**: Assessing the model's performance using metrics such as accuracy, precision, recall, and F1-score.
- **Deployment**: Implementing the trained model into production systems and monitoring its performance.

2.6.1 Machine learning models

1. **Supervised learning**
    - In supervised learning, the algorithm is trained on labeled data, where the input data is paired with the correct output. The goal is

for the model to learn a mapping from inputs to outputs, so that it can predict outputs for new inputs.

2. **Unsupervised learning**
    - Unsupervised learning involves training algorithms on unlabeled data. The model learns patterns and structures in the data without explicit guidance on what to look for. Clustering and dimensionality reduction are common tasks in unsupervised learning.

3. **Semi-supervised learning**
    - Semi-supervised learning leverages both labeled and unlabeled data to improve learning accuracy.
    - This approach is particularly valuable when labeled data is scarce or expensive to obtain, but there is an abundance of unlabeled data available.

4. **Reinforcement learning**
    - RL is a type of learning where an agent learns to make decisions by interacting with an environment. The agent receives feedback in the form of rewards or penalties based on its actions, allowing it to learn optimal behaviors over time [10].

2.6.2 *Applications of machine learning*

- **Natural language processing (NLP)**: ML models are used for language translation, sentiment analysis, text generation, and speech recognition.
- **Computer vision**: ML algorithms power object detection, image classification, facial recognition, and medical image analysis.
- **Recommendation systems**: ML models are used in personalized recommendations for products, movies, music, and content.
- **Predictive analytics**: ML is used in finance for stock market predictions, in healthcare for disease diagnosis and prognosis, and in marketing for customer churn prediction and sales forecasting.
- **Autonomous systems**: ML plays a crucial role in autonomous vehicles, robotics, and automated decision-making systems.

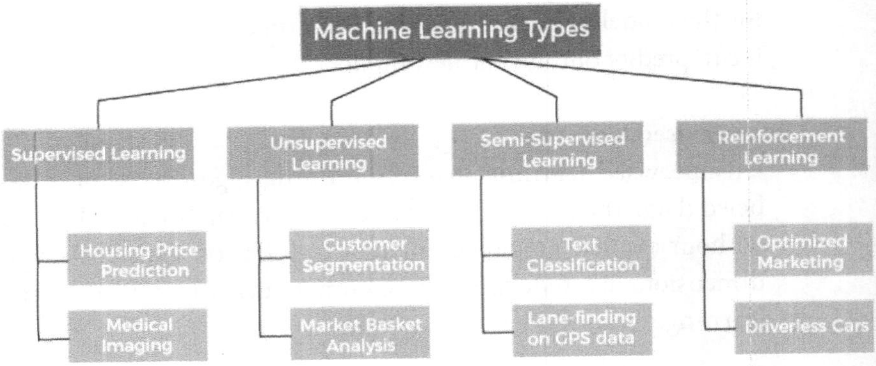

FIGURE 2.4 Machine learning models.

ML plays a significant role in resource allocation across various domains, particularly in intelligent communication networks, cloud computing, and distributed systems (Figure 2.4).

1. Traffic prediction and demand forecasting

   ML models can analyze historical traffic patterns and predict future demand. This capability allows systems to proactively allocate resources based on anticipated workload, thereby optimizing resource utilization and ensuring adequate capacity to handle peak demands without over-provisioning.

2. QoS optimization

   ML algorithms can prioritize resource allocation based on QoS requirements. By learning from past performance metrics and user behavior, ML models can dynamically adjust resource allocations to meet SLAs and ensure consistent QoS for critical applications.

3. Dynamic resource allocation

   ML enables dynamic and adaptive resource allocation by continuously analyzing real-time data such as network traffic, latency metrics, and workload patterns. This allows systems to optimize resource distribution in response to changing conditions, improving overall system efficiency and performance.

4. Anomaly detection and fault management

   ML algorithms can detect anomalies and identify potential fault in resource allocation or usage patterns. By learning normal behavio from historical data, ML models can flag deviations, enabling proactive intervention to prevent service disruptions or optimize resource reallocation in case of failures.

5. Energy efficiency

   ML techniques contribute to optimizing energy consumption by predicting workload patterns and adjusting resource allocations accordingly. This helps in reducing unnecessary power consumption during low-demand periods or by dynamically managing workload distribution across energy-efficient resources.

6. Resource virtualization and allocation

   In virtualized environments like cloud computing, ML enhances resource allocation by intelligently placing workloads on VMs or containers based on performance metrics, cost considerations, and workload characteristics. This optimization minimizes resource contention and improves overall system utilization.

7. Optimization models and algorithms

   ML-driven optimization models, such as RL, GAs, or neural networks, can be applied to solve complex resource allocation problems that involve multiple objectives and constraints. These models can learn from experience and adapt strategies over time to achieve better performance outcomes.

8. Network slicing in 5G networks

   In 5G networks, ML facilitates network slicing by dynamically allocating virtual network resources (e.g., bandwidth, latency) to different slices based on vwancy and personalized service delivery without compromising performance.

2.7 Quality of service

QoS in communication networks refers to the performance level of a service, such as the guarantee of a certain level of bandwidth, latency, jitter, and error rate. QoS is crucial in ensuring that network services meet the performance requirements of various applications and users. QoS is a critical aspect of communication networks, ensuring that various applications and services meet their performance requirements. AI enhances QoS by enabling intelligent, adaptive, and proactive management of network resources. Optimizing resource allocation in intelligent communication networks is essential for maintaining QoS. AI offers powerful tools for dynamically managing network resources, predicting traffic patterns, and adapting to changing conditions to ensure high performance and reliability.

Optimizing resource allocation in intelligent communication networks is crucial for maintaining QoS. Here's a detailed analysis of how QoS principles can be integrated into resource allocation strategies, particularly with the help of AI. While challenges exist in terms of data quality, integration, security, and transparency, the benefits of AI-driven QoS optimization—such as improved efficiency, performance, scalability, and cost reduction—make it a critical approach for modern and future communication networks [5].

2.7.1 Key QoS parameters in resource allocation

1. **Bandwidth**
   Ensuring adequate bandwidth allocation to meet the demands of various applications, such as video streaming and Voice over Internet Protocol (VoIP), which require high throughput.

2. **Latency**
   Minimizing delays to support real-time applications like online gaming and video conferencing, which are highly sensitive to latency.

3. **Jitter**
   Reducing variability in packet arrival times to maintain the quality of audio and video communications.

4. **Packet loss**
   Ensuring low packet loss rates to prevent degradation in service quality, particularly for real-time and critical data transmissions.

2.7.2 Benefits of AI-driven QoS in resource allocation

1. **Enhanced efficiency**
   AI optimizes the use of network resources, reducing waste and ensuring that resources are allocated where they are most needed.

2. **Improved performance**
   By dynamically adjusting to network conditions, AI helps maintain high performance levels, ensuring that QoS requirements are consistently met.

3. **Scalability**
   AI systems can handle the complexity of large-scale networks, making it feasible to manage resource allocation effectively as network demands grow.

4. **Cost reduction**
   Efficient resource allocation reduces operational costs by minimizing over-provisioning and under-utilization of network resources.

5. **Proactive management**
   AI enables proactive management of network resources, preventing QoS degradation before it impacts users.

2.7.3 AI techniques for optimizing resource allocation

1. **Predictive analytics**
   Using ML models to predict network traffic patterns and potential congestion points allows proactive adjustments in resource allocation.

2. **Dynamic resource allocation**
   AI can dynamically allocate resources such as bandwidth, computing power, and storage based on real-time analysis of network conditions and QoS requirements.

3. **Anomaly detection**
   Using AI to detect anomalies in network performance that might indicate issues such as congestion, hardware failures, or security threats enables rapid corrective actions.

4. **Policy management**
   AI systems can optimize QoS policies by learning from historical data and network performance metrics, continually adjusting policies to improve service quality.

5. **Network slicing**
   In 5G networks, AI can manage network slices with different QoS requirements, ensuring each slice receives the necessary resources to meet its specific performance criteria.

2.7.4 AI-driven QoS optimization strategies

1. **Traffic classification and prioritization**
   AI algorithms classify network traffic and prioritize it based on application type, ensuring that high-priority traffic (e.g., emergency

services, real-time communications) receives the necessary resources to maintain QoS.

2. **Adaptive traffic shaping**
   AI dynamically adjusts traffic shaping mechanisms to regulate data flow, ensuring a consistent and predictable traffic pattern that aligns with QoS goals.

3. **Intelligent load balancing**
   AI distributes network traffic across multiple paths or servers to prevent congestion, balancing the load to optimize performance and maintain QoS.

4. **Resource reservation**
   AI can reserve resources for specific applications or services based on their QoS requirements, ensuring that critical services receive the necessary bandwidth and computing power.

2.8 Resource optimization in software-defined networking

Software-defined networking (SDN) offers a flexible and dynamic approach to managing network resources. By decoupling the control plane from the data plane, SDN enables centralized management and programmability, which can be leveraged for resource optimization. Resource optimization in SDN involves leveraging the centralized control and programmability of the network to dynamically manage and allocate resources efficiently. AI and ML play a significant role in enhancing these capabilities by providing predictive insights, anomaly detection, and adaptive policies. Integrating QoS with SDN ensures that the network meets the performance requirements of various applications, leading to improved user experience and efficient utilization of resources (Figure 2.5). Despite the challenges, the combination of SDN, AI, and QoS offers a powerful framework for modern and future communication networks [4].

2.8.1 Key concepts in SDN

SDN introduces several key concepts and resource optimization techniques that enhance network management and performance. One fundamental concept is the separation of the control plane and the data plane, where the control plane decides the traffic routing [11], and the data plane forwards the traffic to its destination. Centralized control through an SDN controller provides a global view of the network, enabling more efficient resource management. SDN's programmability allows network

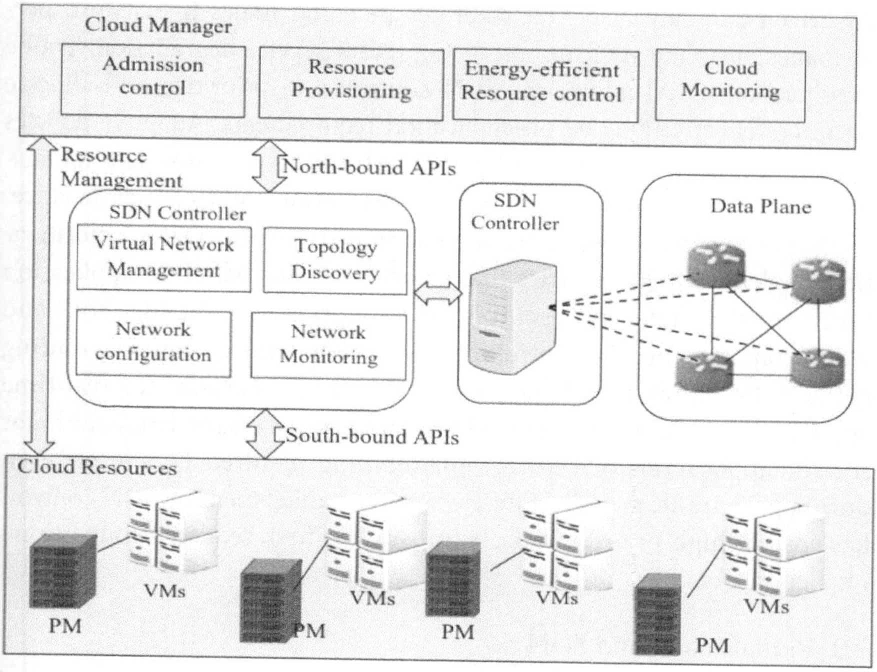

FIGURE 2.5 Architecture of SDN-enabled cloud.

administrators to configure and manage the network programmatically using APIs and high-level programming languages [9].

Resource optimization in SDN is achieved through various techniques. Dynamic traffic engineering allows SDN controllers to adjust traffic flows based on real-time network conditions, optimizing resource use and improving performance. Load balancing, facilitated by monitoring network traffic, distributes loads evenly across network paths to prevent congestion and ensure efficient bandwidth utilization. SDN's fine-grained control over bandwidth allocation ensures that critical applications receive necessary resources to meet QoS requirements. Network slicing in SDN enables the creation of virtual network slices, each with dedicated resources and QoS policies, tailored for different services and applications. Automated provisioning of network resources based on predefined policies and real-time demands reduces manual intervention and operational costs [7].

AI and ML further enhance resource optimization in SDN. Predictive analytics use ML models to forecast future traffic patterns and potential network bottlenecks, allowing proactive resource allocation and optimization.

AI-driven anomaly detection identifies potential issues in network performance, enabling swift responses to prevent service degradation. Traffic classification by AI allows the SDN controller to prioritize and allocate resources effectively based on application requirements. Adaptive policies dynamically adjust network configurations based on real-time data and historical trends, ensuring optimal resource utilization and QoS maintenance.

QoS in SDN is managed through several strategies. Traffic prioritization implements QoS policies that prioritize traffic based on application type, ensuring sufficient resources for high-priority traffic like VoIP and video conferencing. By controlling packet flow and optimizing routing paths, SDN minimizes latency and jitter, which is crucial for real-time applications. Guaranteed bandwidth reserves necessary bandwidth for specific applications or services, maintaining required QoS levels even during peak traffic periods. Effective queue management by SDN controllers ensures high-priority traffic is transmitted first, reducing congestion's impact on QoS [2].

2.9 Virtualization and SDN

Virtualization and SDN are complementary technologies that, when combined, offer powerful capabilities for optimizing resource allocation in intelligent communication networks. Virtualization abstracts physical resources into logical units, while SDN provides centralized control and programmability. Together, they enhance flexibility, efficiency, and scalability in managing network resources.

2.9.1 Benefits of combining virtualization and SDN

Dynamic resource allocation in modern networks leverages virtualization and SDN to optimize resource usage. Virtualization enables the on-demand creation and allocation of VMs or containers for network functions, while SDN dynamically routes traffic to ensure efficient use of these resources. This approach enhances scalability by allowing virtual resources to be easily scaled up or down based on demand. SDN provides the control needed to reconfigure the network dynamically to accommodate these changes. Improved flexibility and agility are achieved through virtualization, which allows rapid deployment and migration of network functions, and SDN, which facilitates seamless reconfiguration of network paths to support these changes. In addition, the cost efficiency of this hybrid technology lies in reducing the need for specialized hardware and optimizing the use of existing resources, thereby lowering capital and operational expenditures

Centralized management capabilities offered by both virtualization and SDN simplify network operations and maintenance [12].

2.9.2 Optimizing resource allocation in the hybrid technology

Optimizing resource allocation in hybrid technologies involves integrating network function virtualization (NFV) and SDN. Dynamic service chaining allows SDN to route traffic through a chain of virtualized network functions (VNFs) based on current network conditions and policies. Resource pooling enables VNFs to be instantiated on any available server within the network, with SDN ensuring traffic is directed to the appropriate VNF instances. Elastic scaling allows VNFs to scale in or out based on real-time traffic demands, with SDN adjusting network paths accordingly. Traffic engineering and load balancing are essential components, where SDN controllers distribute traffic evenly across multiple paths and VNFs, preventing congestion and ensuring efficient resource utilization. Critical services are prioritized through QoS policies, with SDN directing high-priority traffic to optimal paths.

Energy efficiency is also a key focus, achieved through resource consolidation, where virtualization consolidates workloads onto fewer physical servers during low-demand periods, and SDN manages traffic to ensure minimal disruption. Power management allows SDN to direct traffic to active, low-power servers and switch off idle servers to save energy. QoS enforcement by SDN ensures that performance requirements are met for different types of traffic in virtualized environments. Dynamic QoS adjustments, based on real-time monitoring and analytics, allow SDN to adapt QoS settings to changing network conditions and demands. Multi-tenancy support involves virtualization providing isolation between tenants and SDN enforcing security policies to protect tenant traffic. SDN also allocates bandwidth and other resources to different tenants based on their specific needs and SLAs [5].

2.10 Optimization models

Optimization models are crucial for enhancing resource allocation in intelligent communication networks. These models use mathematical and computational techniques to efficiently allocate resources like bandwidth, computing power, and storage to meet the performance requirements of various applications and services. They play a significant role in improving network performance, efficiency, and QoS. Despite challenges related to complexity, scalability, and real-time adaptation, advancements in optimization algorithms, AI, and ML are continually pushing the boundaries of network resource optimization [3].

2.10.1 Key optimization models

Key optimization models include linear programming (LP) and integer linear programming (ILP), which optimize linear objective functions subject to linear constraints. Mixed-integer linear programming (MILP) combines continuous and integer variables, offering flexibility for a range of resource allocation problems. NLP handles more complex scenarios with nonlinear relationships between variables. Stochastic programming addresses optimization under uncertainty, making it suitable for networks with unpredictable traffic patterns. Dynamic programming (DP) solves problems by breaking them into simpler subproblems, effective for sequential decision-making in resource allocation. Game theory models interactions between multiple decision-makers to optimize resource allocation in competitive environments. Heuristic and metaheuristic algorithms, such as genetic algorithms (GA), simulated annealing (SA), and particle swarm optimization (PSO), provide approximate solutions to complex optimization problems. RL, a type of ML, enables agents to learn decision-making by interacting with the environment, useful for real-time resource allocation and adaptation in dynamic networks.

Applications of these optimization models in resource allocation include bandwidth allocation, where LP and ILP optimize distribution to maximize throughput and minimize congestion, while MILP considers both continuous and discrete constraints. Load balancing uses heuristics, metaheuristics, and game theory to distribute traffic evenly across network paths, preventing bottlenecks. Traffic engineering employs stochastic programming and DP to manage network traffic by optimizing routes and resources under variable demand. Network slicing in 5G utilizes ILP, MILP, and RL to allocate resources to virtual network slices, ensuring each meets its QoS requirements. Energy efficiency is optimized using NLP to model nonlinear power usage relationships and heuristics for energy-saving strategies. QoS management uses LP, ILP, and DP to ensure constraints like latency and jitter are met, adaptively managing resources to maintain QoS levels amid changing network [4].

Optimization models play a crucial role in optimizing resource allocation in intelligent communication networks. These models leverage mathematical and computational techniques to make efficient decisions about how resources such as bandwidth, computing power, and storage are allocated to meet the performance requirements of various applications and services Optimization models are essential tools for enhancing resource allocation in intelligent communication networks. By leveraging various mathematical and computational techniques, these models can significantly improve network performance, efficiency, and QoS. While challenges exist in terms of complexity, scalability, and real-time adaptation, ongoing advancements in optimization algorithms, AI, and ML continue to push the boundaries of what is possible in network resource optimization [8].

3 CHALLENGES IN RESOURCE ALLOCATION OPTIMIZATION

3.1 Heterogeneous networks

Integration of diverse network technologies (e.g., 5G, Wi-Fi, satellite) requires adaptive resource allocation strategies that can handle different capabilities and characteristics. Heterogeneous networks (HetNets) play a crucial role in optimizing resource allocation in intelligent communication networks.

3.1.1 Challenges in HetNets

1. **Interference management**

   Challenge: Managing interference between different types of cells and technologies.

   Solution: Advanced interference coordination techniques such as Coordinated MultiPoint (CoMP), Enhanced Inter Cell Interference Coordination (eICIC), and beamforming.

2. **Handover management**

   Challenge: Ensuring seamless handovers between different cells and technologies.

   Solution: Improved handover algorithms and signaling protocols to minimize latency and packet loss.

3. **Backhaul connectivity**

   Challenge: Providing robust and scalable backhaul connections for small cells.

   Solution: Utilizing high-capacity backhaul solutions such as fiber, microwave, and millimeter-wave technologies.

4. **Resource allocation complexity**

   Challenge: Optimizing resource allocation in a highly heterogeneous and dynamic environment.

Solution: AI and ML algorithms for dynamic and adaptive resource management.

5. **Security and privacy**

   Challenge: Ensuring secure communication and protecting user data across diverse network components.

   Solution: Implementing robust security protocols and encryption techniques.

3.2 Security and privacy

Optimizing resource allocation in intelligent communication networks requires addressing significant security and privacy challenges. These challenges stem from the need to protect sensitive data, ensure the integrity and availability of network services, and comply with regulatory requirements, while efficiently managing resources. Optimizing resource allocation in intelligent communication networks involves addressing complex security and privacy challenges. Solutions require a holistic approach that integrates robust security measures, compliance with privacy regulations, and efficient resource management strategies. Leveraging advanced technologies such as AI, ML, and blockchain can enhance both security and efficiency, ensuring that network resources are allocated optimally without compromising on security and privacy [13].

3.2.1 Challenges in security and privacy

1. **Data confidentiality and integrity**

   Eavesdropping and interception: Sensitive data can be intercepted during transmission, leading to breaches of confidentiality.

   Data tampering: Ensuring that data remains unaltered during transmission and storage is crucial to maintain its integrity.

2. **Authentication and authorization**

   Unauthorized access: Ensuring that only authorized users and devices can access network resources.

   Identity spoofing: Preventing malicious actors from impersonating legitimate users or devices.

3. **Availability and resilience**

   Denial-of-service (DoS) attacks: Protecting the network from attacks that aim to disrupt service availability.

   Resource exhaustion: Preventing attackers from consuming excessive network resources, leading to performance degradation.

4. **Privacy compliance**

   Regulatory requirements: Complying with privacy regulations such as General Data Protection Regulation (GDPR), California Consumer Privacy (CCPA), and others.

   Data minimization: Ensuring that only necessary data is collected and processed, in line with privacy principles.

5. **Interference and jamming**

   Signal interference: Protecting the network from intentional or unintentional interference that can disrupt communication.

   Jamming attacks: Preventing malicious entities from jamming network signals to disrupt service.

6. **Secure and efficient resource allocation**

   Balancing security and efficiency: Ensuring that security measures do not overly compromise network performance and resource allocation efficiency.

   Dynamic threat landscape: Adapting to constantly evolving security threats that can affect resource allocation strategies.

7. **Secure communication protocols**

   Protocol vulnerabilities: Ensuring that communication protocols used for resource allocation are secure and free from vulnerabilities.

   Backward compatibility: Maintaining security while ensuring compatibility with older devices and protocols.

8. **End-to-end security in distributed environments**

   Distributed architectures: Ensuring consistent security across distributed network components, such as edge devices and cloud servers.

Data privacy in multi-tenant environments: Protecting data privacy in shared or multi-tenant environments.

9. **Energy efficiency and security trade-offs**

   Resource constraints: Balancing the need for energy efficiency with the implementation of robust security measures.

   Performance overheads: Minimizing the performance overhead introduced by security mechanisms, which can affect resource allocation efficiency.

4 FUTURE DIRECTIONS

4.1 5G and beyond

The advent of 5G and the upcoming 6G technologies represent significant advancements in optimizing resource allocation within intelligent communication networks. These technologies introduce features and capabilities that address the growing demands for bandwidth, latency, and connectivity, making resource allocation more efficient and effective. As we look toward the future, 6G is driven by the need to meet ever-increasing demands for higher data rates, lower latencies, enhanced connectivity, and improved user experiences.

Terahertz (THz) communication, operating at frequencies between 0.1 and 10 THz, offers ultrahigh data rates and is suitable for short-range applications due to higher propagation losses. This makes THz communication ideal for indoor environments and dense urban areas. Future directions include the development of new modulation techniques and innovations in THz transceiver design to efficiently handle high-frequency signals.

AI and ML integration will play a critical role in automating network management and optimization processes. AI can be utilized for predictive maintenance and network performance forecasting. Future directions involve the development of AI-driven network management systems that can adapt in real-time to changing conditions and the use of RL to continuously improve resource allocation and QoS.

Quantum communication and computing promise advancements such as quantum key distribution (QKD) for unbreakable encryption and potential low-latency communication using quantum entanglement. Future research is focused on integrating quantum communication with classical networks to create a hybrid quantum internet and implementing quantum encryption protocols in communication networks to enhance security.

Integrated sensing and communication (ISAC) combines communication and sensing capabilities within the same infrastructure, enabling applications like environmental monitoring and smart cities. Future developments include systems that can simultaneously perform radar sensing and communication, as well as the integration of ISAC in urban infrastructure for real-time monitoring and management.

Advanced antenna technologies, such as massive MIMO and intelligent reflecting surfaces (IRS), enhance capacity and coverage by dynamically controlling electromagnetic wave propagation. Future directions explore holographic MIMO to enhance spatial multiplexing capabilities and adaptive beamforming for optimizing performance in dynamic environments.

Enhanced edge computing brings computational resources closer to end-users to minimize latency and enable real-time analytics and decision-making. Future directions involve implementing federated learning at the edge to train AI models across distributed devices without centralizing data and developing AI algorithms specifically optimized for edge computing environments.

Advanced network architectures include network slicing to create virtualized network slices tailored to specific use cases and decentralized architectures to improve resilience and scalability. Future developments focus on real-time dynamic network slicing to adapt to changing demands and the use of blockchain technology to enhance security, transparency, and trust in decentralized network management.

Sustainability and green networking aim to reduce energy consumption and integrate renewable energy sources within network infrastructure. Future directions involve developing energy harvesting technologies that enable network devices to harvest energy from their environment and implementing measures to minimize the carbon footprint of network operations through efficient design and operation.

Enhanced user experience and connectivity ensure seamless connectivity across different network types and environments, supporting applications like augmented reality (AR), virtual reality (VR), and extended reality (XR) with high bandwidth and low latency. Future directions explore new use cases for 6G, such as holographic communications, tactile internet, and multi-sensory experiences, as well as achieving universal connectivity in remote and underserved areas through innovative deployment strategies and satellite communication integration [6].

5 CONCLUSION

Semi-supervised learning represents a powerful approach in ML, enabling the use of vast amounts of unlabeled data to build more accurate and efficient models. By leveraging techniques such as self-training, co-training, graph-based methods, and generative models, Secure Sockets Layer (SSL) can be applied across various domains, including NLP, computer vision, healthcare, finance, and speech recognition. Despite its challenges, such as label propagation errors and scalability issues, SSL continues to be an area of active research and significant potential, particularly in data-rich environments where labeled data is scarce.

Optimizing resource allocation in intelligent communication networks presents a multifaceted challenge requiring a nuanced approach. By integrating advanced algorithms and AI-driven solutions, networks can efficiently distribute resources, enhancing performance, reliability, and scalability. However, achieving optimal allocation demands addressing complexities such as dynamic traffic patterns, diverse user demands, and evolving technology landscapes. Embracing these challenges with innovative strategies and collaborative efforts among stakeholders will pave the way for robust, adaptive networks capable of meeting future communication demands effectively.

REFERENCES

1. H. Nalband, M. Sarvagya, and M. Riyaz, "Spectral efficient beamforming for mmWave MISO systems using deep learning techniques," *Arab. J. Sci. Eng.*, vol. 46, no. 10, pp. 9783–9795, 2021, doi: 10.1007/s13369-021-05552-4.
2. D. Zhang, Y. Fang, Y. Zhou, J. He, and Y. Zhang, "Game theoretic multihop D2D content sharing: Joint participants selection, routing, and pricing," IEEE Transactions on Mobile Computing, p. 143, 2020.
3. H. Houssein Essam, B. El-din Helmy, D. Oliva, A. A. Elngar, and H. Shaban, "A novel Black Widow Optimization algorithm for multilevel thresholding image segmentation," Expert Systems with Applications, vol. 167, p. 14159, 2021.
4. M. I. Khan, L. Reggiani, M. M. Alam, Y. Le Moullec, N. Sharma, E. Yaacoub, and M. Magarini, "Q-learning based joint energy-spectral efficiency optimization in multi-hop device to-device communication," Sensors, vol. 20, no. 22, p. 6692, 2020.
5. P. K. Malik, D. S. Wadhwa, and J. S. Khinda, "A survey of device to device and cooperative communication for the future cellular networks," International Journal of Wireless Information Networks, vol. 27, no. 3, pp. 411–432, 2020
6. S. M. Hosseini, S. Shahsavari, and C. Rosenberg, "Multi-user scheduling in hybrid millimeter wave massive MIMO systems," Sep. 2021, [Online] Available: https://arxiv.org/abs/2109.05681
7. K. E. Purushothaman and V. Nagarajan, "Multi-objective optimization based on self-organizing PSO algorithm for massive MIMO 5G

wireless network," International Journal of Communication Systems, ISSN: 1074-5351, vol. 34, no. 4, pp. 1-15, 2021.

8. P. R. Teja and P. K. Mishra, "Path selection and resource allocation for 5G multi-hop D2D networks," Computer Communications, vol. 195, pp. 292–302, 2022.
9. T. P. Lillicrap, J. J. Hunt, A. Pritzel, N. M. O. Heess, T. Erez, Y. Tassa, D. Silver, and D. P. Wierstra, "Continuous control with deep reinforcement learning," Sep. 15 2020, US Patent 10,776,692.
10. X. Shang, H. Yin, Y. Wang, M. Li, and Y. Wang, "Secure multiuser scheduling for hybrid relay-assisted wireless powered cooperative communication networks with full-duplex destination-based jamming," IEEE Access, vol. 9, pp. 49774–49787, 2021. doi: 10.1109/ACCESS.2021.3067472.
11. S. Sharma and V. K. Verma, "AIEMLA: Artificial intelligence enabled machine learning approach for routing attacks on internet of things," The Journal of Supercomputing, vol. 77, no. 12, pp. 13757-13787, 2021, doi: 10.1007/s11227-021-03833-1.
12. J. Anjana Devi and V. Prasanna Srinivasan, "A two-fold optimization framework using hybrid B2 algorithm for resource allocation in long-term evolution based cognitive radio networks system," International Journal of Communication System, vol. 35, no. 3, pp. 1–24, 2022.
13. P. R. Teja and P. K. Mishra, "Sealed bid single price auction model (SBSPAM)-based resource allocation for 5G networks," Wireless Personal Communications, vol. 116, no. 3, pp. 2633–2650, 2021.

CHAPTER 3

Radio resource management for M2M communications in cellular networks

Mariyam Ouaissa and Mariya Ouaissa

1 INTRODUCTION

Machine-to-machine (M2M) communication is an emerging form of communication that will soon be integrated into our current high-speed mobile networks, including long-term evolution (LTE) and beyond. Unlike traditional human-to-human (H2H) communication, M2M communication is characterized by significant signaling and small data packet sizes [1]. These unique features introduce new challenges for LTE networks, such as managing radio access for a large number of devices and allocating radio resources while maintaining quality of service (QoS) through effective scheduling mechanisms [2]. An efficient scheduling algorithm must ensure desired QoS, guaranteeing throughput, delay, and fairness, while optimizing the use of available radio resources. Schedulers range from simple to complex designs, tailored to various QoS requirements [3].

This chapter focuses on resource management and sharing within the LTE access network. Different protocols are used for this purpose;

DOI: 10.1201/9781003514336-3

some utilize contention-based random access (RA) for resource allocation, while others use fixed allocation, known as scheduling techniques. Section 2 describes the architecture and structure of the LTE radio interface, the functions of cellular network channels, and the resource management mechanisms, providing an overview of LTE network QoS. The random-access channel (RACH) is particularly relevant and is detailed in terms of its structure, function, and access protocol in Section 3. Section 4 delves into the principles of uplink and downlink scheduling.

2 STRUCTURE AND MANAGEMENT OF THE LTE RADIO INTERFACE

The radio interface plays a crucial role in transmitting data over the air from the IP layer, which is linked to the user's requested service. This transmission must adhere to QoS requirements such as latency and throughput, despite the highly variable nature of the medium. In addition, optimizing access to limited spectral resources and adapting to the availability of spectrum, which varies globally, is essential [4].

2.1 Architecture and structure of the radio interface over LTE

2.1.1 Layered architecture of the LTE air interface

The LTE interface consists of both vertical and horizontal layers. The horizontal layers follow the OSI model, while the vertical layers divide the interface into the control plane and the user plane. The radio interface ensures reliable connections between user equipment (UE) and eNodeBs. It comprises three layers: the physical layer, the link layer, and the radio resource control (RRC) layer [5].

2.1.2 Radio interface protocol architecture

The radio interface enables data transfer between the UE and the eNodeB while meeting QoS requirements and optimizing spectral resource access. The protocol stack facilitates the exchange of data and signaling messages within the EPS network. The LTE protocol stack includes two planes: the user plane, which manages the routing of useful data, and the control plane, which handles the management of signaling messages. The transport protocols of the air interface are grouped into two layers: the physical layer, responsible for digital and analog signal processing, and the data link layer, which includes the MAC, RLC, and PDCP protocols [6].

2.1.3 LTE radio interface communication channels

The LTE system uses channels to identify data types transported on the radio interface and the physical parameters related to transmission. The radio interface channels are service access points provided by an N layer. Each sub-layer receives two data streams from the upper sub-layer for each active UE: one IP data stream and one RRC control flow. These packets are processed and sent as PDUs to the directly lower sub-layer. The interactions between sub-layers are facilitated by communication channels for packet transfer.

2.1.4 Control information exchanged over the LTE radio interface

Control information exchanged between an eNodeB base station (BS) and active UEs can be classified into four categories: uplink transmissions, downlink transmissions, format control indicators, and hybrid ARQ acknowledgments.

2.1.4.1 Control information transmitted in uplink mode

- **Channel quality indicator (CQI)**: Indicates the maximum data rate an UE can handle with a block error rate (BLER) of ≤10%. The CQI value, ranging from 0 to 15, is influenced by the signal-to-interference-plus-noise ratio (SINR) and the UE's performance.
- **Rank indicator (RI)**: Optimizes and maximizes the capacity of the downlink radio channel by indicating the maximum number of different signals a downlink user can receive simultaneously. This depends on the number of antennas used by the eNodeB.
- **Precoding matrix indicator (PMI)**: Specifies the precoding matrix the eNodeB should apply to data streams before transmission maximizing the number of data bits received by the UE through all antennas.

2.1.4.2 Other uplink control information

- **Scheduling requests (SR)**: Triggered when the UE has data awaiting transmission on the Physical Uplink Control Channel (PUSCH).
- **Hybrid ARQ acknowledgments**: Sent on the uplink in response to transmissions from the eNodeB to the UE on the shared downlink channel (DL-SCH).

2.1.4.3 Control information transmitted in downlink mode

- **Scheduling commands**: The eNodeB alerts the UE about upcoming transmissions on the Physical Downlink Shared Channel (PDSCH).
- **Uplink scheduling privileges**: The eNodeB grants the UE transmission privileges on the uplink.
- **Transmission capacity control commands in uplink**: The eNodeB adjusts the power with which UEs transmit using these commands.
- **Format control indicator**: Informs UEs about the organization of data and control information on the downlink.
- **Hybrid ARQ indicator**: Acknowledgments sent on the downlink in response to UE transmissions on the Uplink Shared Channel (UL-SCH).

2.2 Radio resource management

In cellular networks, telecommunications operators focus on efficiently utilizing the available bandwidth. Effective exploitation of radio resources enables servicing a larger number of users and reduces infrastructure operating costs by minimizing the number of nodes, such as BSs or eNodeBs. For users, the demand for new services with higher quality remains essential. To meet both operators' and users' needs, a robust management mechanism is necessary [7]. A key feature of the eNodeB BS is the management of radio resource management (RRM), which involves procedures aimed at optimizing the use of radio resources within a cell while satisfying QoS requirements, maximizing throughput, and improving fairness among UEs. RRM encompasses four fundamental mechanisms [8].

1. **Admission control**: Ensures the continuity of existing services while accommodating new connection requests.
2. **Power control**: Extends battery life through optimal signal power management during transmission and reception.
3. **Handover control**: Allows users to maintain service continuity while moving.
4. **Packet scheduling**: Allocates and shares radio resources among active users in a cell sector, based on criteria such as UE waiting time, channel quality, throughput, and fairness.

2.2.1 Admission control

Admission control is crucial for regulating network operations to guarantee service while accommodating new connections optimally. The decision to admit or reject a new user is based on factors like the additional interference the user may cause, resource availability, impact on current QoS, and user priority. Admission control may block new connections if they are likely to degrade ongoing connections' quality or if the requested connection quality cannot be satisfied. This mechanism also prevents congestion of existing connections, ensuring that the radio interface can maintain a satisfactory QoS for all users. For instance, blocking a new connection may be preferable to risking the loss of an already established connection.

In fourth-generation networks, diverse services such as video conferencing, online gaming, and Internet Protocol Television (IPTV) necessitate new admission control schemes to support various QoS profiles. The 3rd Generation Partnership Project (3GPP) group does not specify admission control algorithms; they are vendor-specific.

2.2.2 Link adaptation and scheduling

Link adaptation involves techniques to adjust transmission parameters to instantaneous radio link changes, including adaptive selection of the modulation and coding scheme (MCS). The scheduler performs resource allocation based on physical layer measurements. In dynamic scheduling, link adaptation occurs at each subframe served to the UE, while in semi-persistent scheduling, it remains constant across subframes. The eNodeB calculates uplink channel quality from the UE's transmitted SRS reference signals but cannot measure the downlink channel quality experienced by the UE directly. These indicators depend on received power, interference levels on the physical resource block (PRB), and the UE receiver.

Scheduling is vital for allocating radio resources to users within a cell. Scheduling algorithms and link adaptation require CQI values as deterministic information of channel quality. Link adaptation algorithms also need scheduler decisions to determine expected users and assigned resource blocks (RBs). The results of these algorithms are communicated to the UEs via the PDCCH channel. By effectively managing these aspects, LTE networks can ensure efficient use of radio resources, maintain high QoS, and support a wide range of user services [9].

3 RANDOM ACCESS FOR M2M COMMUNICATIONS IN LTE

A fundamental requirement for all cellular systems is the ability for terminals to request the establishment of a connection. This process, known as the RA procedure, is essential for initial access, transitions from inactive to active states, and reestablishing activity after periods of inactivity in the uplink direction. When a terminal is inactive, no radio resources are dedicated to it. Therefore, to initiate contact with the eNodeB, the terminal must use resources accessible to all terminals. This necessitates a method to limit collisions—simultaneous transmissions by multiple terminals on the same resources—and resolve them if they occur. The RACH is the initial channel through which a user connects to a cellular network. It is an uplink channel that transmits control information to the BS or eNodeB using a RA technique. Congestion at the RACH, caused by simultaneous access attempts by numerous devices, can significantly impact system performance. Thus, this section focuses on the RACH, a primary challenge in supporting M2M traffic over cellular networks [10].

3.1 The random-access procedure in LTE

In the RA procedure, the terminal selects a number called a preamble. The terminal, which can be an UE or an MTC device (MTCD), either randomly selects this preamble from a given set or receives it from the eNodeB via RRC signaling, in which case it is known as a dedicated preamble. Each terminal transmits the necessary information using allocated resources, and the eNodeB confirms the procedure's success to the terminal. Once this confirmation is received, the eNodeB and the terminal can initiate data communication. The LTE system uses two types of RA models: contention-free and contention-based. This work focuses on the contention-based RA model in LTE.

3.1.1 Random access without contention

In contention-free mode, the eNodeB assigns a distinct preamble to each terminal, eliminating collision problems (Figure 3.1). This mode is used in scenarios requiring low latency, such as downlink traffic forwarding and resumption. In these cases, a dedicated preamble is assigned as needed. The contention-free procedure is also utilized in handover scenarios, which include:

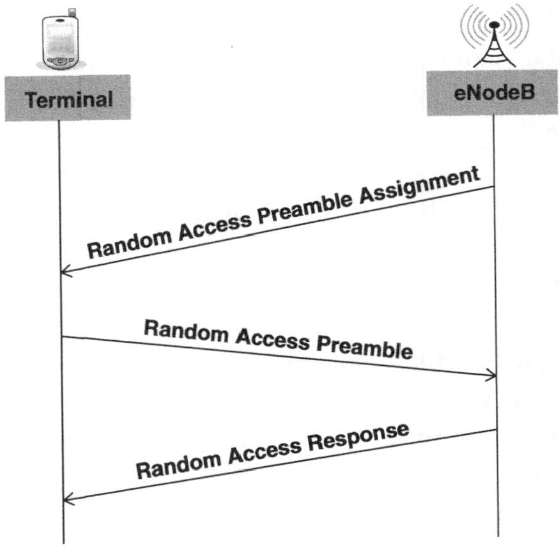

FIGURE 3.1 The contention-free random-access procedure.

Intra radio access technology (Intra-RAT): Handover within the same radio access technology, such as LTE to LTE from one eNodeB to another.

Inter radio access technology (Inter-RAT): Handover between different radio access technologies, such as between LTE and GSM, UMTS, WiMAX, or WLAN.

3.1.2 The contention-based random-access procedure

The contention-based RA procedure consists of four steps (Figure 3.2):

Phase 1: Random-access preamble

In the first stage, the terminal emits an RA preamble, allowing the eNodeB to estimate the terminal's transmission timing. Synchronization is essential; without it, the terminal cannot transmit data upstream. The terminal sends a randomly chosen preamble in an RA slot and waits for a response from the eNodeB. If the current time slot is not an RA slot, the terminal waits for the next one.

Phase 2: Random-access response

In the second stage, the eNodeB sends a message to adjust the terminal's transmission synchronization based on measurements from the first stage. After sending the preamble, the terminal

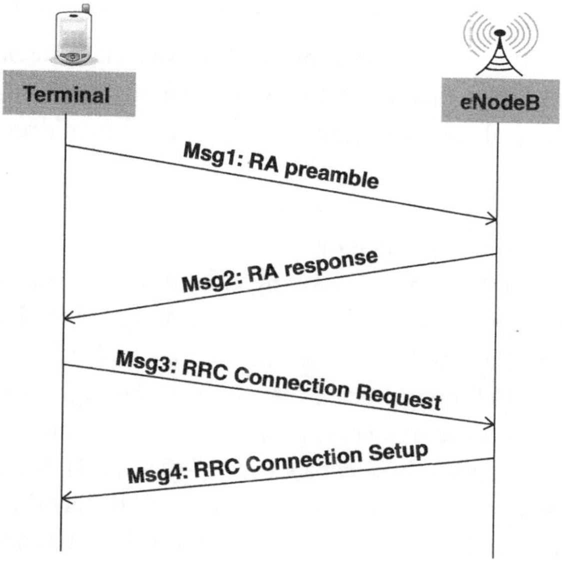

FIGURE 3.2 The contention-based random-access procedure.

increases its transmission count by one, starts the random-access response (RAR) timer, and waits for the RA response from the eNodeB.

Phase 3: RRC connection request

In the third stage, the terminal transmits its identity to the network using the UL-SCH channel. Upon receiving the RAR within the RAR window, the terminal prepares to send the RRC connection request. If no response is received in the RAR window—indicating the eNodeB did not detect the preamble, possibly due to collisions—the terminal retransmits Msg3, causing potential collisions if multiple terminals choose the same preamble.

Phase 4: RRC connection setup

In the fourth stage, the network sends a contention resolution message to the terminal via the DL-SCH channel. If the terminal does not receive this message, it declares a contention resolution failure and plans a new access attempt, restarting the process with a new preamble. Each device maintains a preamble transmission counter, which increases with each unsuccessful attempt. When the counter reaches its maximum value, the network is declared unavailable, and an RA problem is indicated to the upper layers.

3.2 3GPP random-access mechanisms for M2M communications

Research in 3GPP indicates that both MTC and H2H devices will experience continuous collisions on the RACH when numerous MTCDs are introduced. Several solutions have been proposed to address this congestion problem:

3.2.1 Access class barring scheme

In UTRAN/E-UTRAN, access barring effectively controls MTCD access, especially during network overload. Initially designed for H2H terminals, access class barring (ACB) scheme uses 16 classes (AC = 0–9 for normal terminals, AC = 10 for urgent cases, AC = 11–15 for high-priority services). The eNodeB broadcasts an access probability "p" and an AC barring time delay. Terminals randomly select a value "q" ($0 \leq q \leq 1$). If $q \leq p$, the terminal proceeds with the RACH procedure; if $q > p$, the terminal waits for the AC barring time before retrying. ACB can control MTCD access using low p-values, although this increases access timeout values. An MTC-specific access class could further improve network control granularity.

3.2.2 Separation of RACH resources

To prevent congestion and maintain conventional communications, RACH resources can be separated for H2H and MTCDs. This can be done by dividing preambles into two groups or by separating time-frequency resources of the RACH channel (RBs) for H2H and MTCDs. Two methods are detailed: one separates the available preambles into disjoint subsets; and the other divides preambles into subsets for H2H clients and shared subsets for both H2H and MTC clients.

3.2.3 Dynamic resource allocation RACH

This solution allows the network to dynamically allocate additional RACH resources when anticipating overload due to excessive MTCD access attempts. While effective in some cases, it is limited by the availability of additional resources.

3.2.4 Backoff specific scheme

This mechanism assigns a shorter timeout duration to H2H terminal than to MTCDs, reducing collision and congestion in the access network. However, it may cause delays that negatively impact high-priority applications. While effective with low congestion, it is less so with high congestion levels on the RACH channel.

3.2.5 Slotted access

Access slots are designated for MTCDs, allowing network access only at the start of their dedicated time slot. This approach prevents MTCDs from accessing the network at will, reducing congestion at the access level.

4 SCHEDULING OF M2M COMMUNICATIONS IN LTE NETWORKS

RRM is a critical function of BSs in cellular systems. This function is paramount in LTE systems to ensure QoS, maximize capacity, and improve fairness among devices vying for radio resources. However, these goals often conflict; enhancing one aspect may degrade another. For instance, increasing a cell's overall throughput might reduce fairness between terminals or decrease QoS. RRM techniques strive to optimize radio resource usage while meeting QoS requirements for as many users as possible, despite challenges like limited spectrum, signal distortion from multipath channels, user mobility, and traffic heterogeneity. The scheduling process, managed by eNodeB protocols at the MAC sublayer, determines how radio interface resources are allocated and used in both downlink and uplink modes [11].

4.1 Scheduling process for M2M communications

LTE scheduling techniques are classified based on M2M communication requirements for optimal integration.

1. **Energy-efficient scheduling**: Utilizes single-carrier frequency-division multiple access (SC-FDMA) for the uplink and orthogonal frequency-division multiple access (OFDMA) for the downlink.
2. **QoS-based scheduling**: Manages different QoS requirements such as latency, jitter, error rate, and guaranteed bit rate (GBR).
3. **Multi-hop-based scheduling**: Uses multiple hops to transmit data over short distances, saving power compared to long-distance single-hop transmissions.
4. **Scalable network scheduling**: Supports a vast number of M2M communications using LTE's native IP connectivity feature.

The eNodeB allocates radio resources to user equipment or MTCDs in the downlink or uplink direction. The smallest unit of radio resources assignable to an UE/MTCD is the PRB, comprising 12 subcarriers in a time slot. The scheduling process involves two steps:

Time domain packet scheduling (TDPS): The eNodeB selects a terminal (UE or MTCD) or a group of terminals to allocate PRBs based on criteria like channel state, QoS, and fairness.

Frequency domain packet scheduling (FDPS): The eNodeB assigns PRBs to the terminal(s) selected in the TDPS step, aiming to maximize their use based on channel conditions.

Designing scheduling techniques for M2M communications must consider MTCDs' stringent delay requirements and the competition for limited radio resources among many MTCDs. The scheduler, a key MAC layer function in LTE eNodeB, significantly impacts system performance by determining resource allocation to maximize throughput while ensuring packet delay and QoS guarantees.

4.2 Characteristics of radio resources

Scheduling optimizes RRM and allocation, translating to radio spectrum allocation as channels based on criteria like channel state, available RBs, and QoS constraints. LTE transmissions are organized into 10 ms radio frames, divided into ten 1 ms subframes. Each subframe splits into two 0.5 ms slices. Transmissions use SC-FDMA for the uplink and OFDMA for the downlink. Resources are distributed among different channels for data and control messages [12].

The PDSCH sends data in the downlink direction, while PUSCH sends data in the uplink direction. Control packets on the PDCCH carry terminal identification and data transmission instructions. The PUCCH transports terminal messages containing channel quality information, acknowledgments (ACK), and SRs to the eNodeB. The BSR (buffer status report) and SR messages help terminals verify uplink resource transmission on the PUSCH.

Scheduling decisions are made every 1 ms transmission time interval (TTI) based on algorithms considering user channel condition, interference, QoS criteria, service priorities, and so on. A radio RB corresponds to 12 subcarriers of 15 kHz each (180 kHz) in the frequency domain and a 0.5 ms time slot in the time domain, each slot comprising six or seven OFDM symbols depending on the cyclic prefix (normal or extended).

4.3 Downlink resource allocation

4.3.1 Downlink scheduling procedure

Scheduling allows the eNodeB to allocate physical resources to terminals (UE or MTCD) per algorithm, enabling them to send or receive data in each TTI. Downlink scheduling involves allocating RBs and MCS to each user based on CQIs and other scheduler information. Users are informed about resource allocation via allocation reports to decode data correctly.

4.3.2 Downlink scheduling algorithms

Packet scheduling mechanisms optimize RRM in cellular networks, considering transmission channel state and user QoS requirements, especially for real-time flows like VoIP and video. Radio resource allocation algorithms aim to enhance system performance by improving spectral efficiency and network fairness, balancing efficiency and equity among users [13]. Key algorithms include:

a. **Proportional fair (PF)**: An opportunistic algorithm using infinite queues for non-real-time traffic, aiming to maximize overall throughput while ensuring fairness.

b. **Exponential proportional fair (EXP-PF)**: An enhanced PF algorithm prioritizing real-time streams (multimedia) over others.

c. **Maximum largest weighted delay first (M-LWDF)**: Time-sensitive algorithm for real-time flows (multimedia and VoIP), removing packets exceeding delay tolerances to maintain QoS.

d. **Frame level scheduler (FLS)**: Multi-class scheduling algorithm considering stream types (real-time or non-real-time) for appropriate RB allocation, balancing system throughput and fairness.

e. **Exponential rule (EXP-Rule)**: Improved EXP-PF scheduler, providing QoS guarantees over wireless links by selecting a single user per queue at each scheduling instant.

f. **Logarithmic rule (LOG-Rule)**: Scheduler balancing QoS parameters in terms of average delay, using logarithmic delay calculations.

4.4 Uplink resource allocation

4.4.1 Uplink scheduling procedure

Upstream access uses SC-FDMA, similar to OFDMA in capacity but with a lower crest factor, reducing power requirements and suiting mobile equipment's limited battery life. SC-FDMA's constraint of contiguous frequency domain resource allocation simplifies resource announcement. Uplink data transmission refers to data sent from devices to the network.

Terminals initiate data transfer by sending an SR on PUCCH. The eNodeB responds with DCI 0 assignment on PDCCH, enabling terminals to transmit data on PUSCH based on UL assignment specified by DCI 0. HARQ feedback (ACK/NACK) on PHICH follows; NACK prompts data retransmission by the terminal [14].

4.4.2 Uplink resource allocation techniques

Various algorithms have been proposed to efficiently allocate resources, considering traffic type and user QoS requirements:

a. **Round robin (RR)**: Classic scheduler distributing RBs among all terminals requesting radio resources, ensuring equal packet transmission time sharing but degrading throughput due to lack of SINR-based scheduling.

b. **First maximum expansion (FME)**: Algorithm allocating resources starting from the matrix's highest value, expanding allocation based on adjacent RB values.

c. **Maximum throughput (MT)**: Allocates RBs to UEs with the best radio channel quality, maximizing overall network throughput.

5 CONCLUSION

In this chapter, we provided an overview of resource management in LTE cellular networks, focusing on the two main areas of RA and scheduling. First, we discussed the RACH, emphasizing its crucial role in cellular networks and the necessity for overload protection to prevent network performance degradation from congestion and simultaneous access attempts by multiple devices. Second, we explored the scheduling mechanisms for both downlink and uplink, detailing the operating principles of various resource allocation techniques described in the literature.

REFERENCES

1. B. Agarwal, M. A. Togou, M. Marco, and G. M. Muntean, "A comprehensive survey on radio resource management in 5G HetNets: Current solutions, future trends and open issues," *IEEE Communications Surveys & Tutorials*, vol. 24, no. 4, 2022, pp. 2495–2534.
2. S. Manap, K. Dimyati, M. N. Hindia, M. S. A. Talip, and R. Tafazolli, "Survey of radio resource management in 5G heterogeneous networks," IEEE Access, vol. 8, 2020, pp. 131202–131223.

3. A. Chernov, V. Matyukhin, M. Somov, E. Barashov, J. Chernova, and V. Mishin, "Radio resource management in 5G networks," In *2020 International Conference Engineering and Telecommunication (En&T)*, Dolgoprudny, Russia, pp. 1–5, IEEE, 2020.
4. M. Ouaissa, A. Rhattoy, and I. Chana, "New security level of authentication and key agreement protocol for the IoT on LTE mobile networks," In *2018 6th International Conference on Wireless Networks and Mobile Communications (WINCOM)*, Marrakesh, Morocco, pp. 1–6, IEEE, 2018.
5. M. Ouaissa and A. Rhattoy, "A secure model for machine to machine device domain based group in a smart city architecture," International Journal of Intelligent Engineering & Systems, vol. 12, no. 1, 2019, pp. 151–163.
6. M. J. Alam, M. R. Hossain, S. Azad, and R. Chugh, "An overview of LTE/LTE-A heterogeneous networks for 5G and beyond," Transactions on Emerging Telecommunications Technologies, vol. 34, no. 8, 2023, p. e4806.
7. P. Goyal, V. Rishiwal, and A. Negi, "A comprehensive survey on QoS for video transmission in heterogeneous mobile ad hoc network," Transactions on Emerging Telecommunications Technologies, vol. 34, no. 7, 2023, p. e4775.
8. N. Rajule, M. Venkatesan, R. Menon, and A. Kulkarni, "Mobility prediction in cellular networks: A survey," In *2023 International Conference on Recent Trends in Electronics and Communication (ICRTEC)*, Mysore, India, pp. 1–9, IEEE, 2023.
9. I. Ahmad, M. G. AlFailakawi, A. AlMutawa, and L. Alsalman, "Container scheduling techniques: A survey and assessment," Journal of King Saud University-Computer and Information Sciences, vol. 34, no. 7, 2022, pp. 3934–3947.
10. M. Ouaissa, M. Benmoussa, A. Rhattoy, M. Lahmer, and I. Chana, "Performance analysis of random access mechanisms for machine type communications in LTE networks," In *2016 International Conference on Advanced Communication Systems and Information Security (ACOSIS)*, Marrakesh, Morocco, pp. 1–6, IEEE, 2016.
11. S. Tang, Y. Yu, H. Wang, G. Wang, W. Chen, Z. Xu, S. Guo, W. Gao, W. (2023). "A survey on scheduling techniques in computing and network convergence," IEEE Communications Surveys & Tutorials, vol. 26, no. 1, 2023, pp. 160–195.
12. M. Kumar, S. C. Sharma, A. Goel, and S. P. Singh, "A comprehensive survey for scheduling techniques in cloud computing," Journal of Network and Computer Applications, vol. 143, 2019, pp. 1–33.
13. M. Ouaissa and A. Rhattoy, "New method based on priority of heterogeneous traffic for scheduling techniques in M2M communications over LTE networks," International Journal of Intelligent Engineering & Systems, vol. 11, no. 6, 2018, pp. 209–219.
14. M. Ouaissa and A Rhattoy, "QoS hybrid uplink scheduler based on service type for M2M communications in LTE networks," *Indonesian Journal of Electrical Engineering and Computer Science*, vol. 14, no. 3, 2019, p. 1460.

CHAPTER 4

Integrating blockchain for secure and efficient radio resource management in 5G and beyond networks

D. Suganya, S. Sri Saye Lakshmi,
R. Jayalakshmi and V. Kavitha

1 INTRODUCTION

The rapid evolution of wireless communication technologies has brought about significant advancements and challenges. As the demand for higher data rates, ultralow latency, and massive device connectivity continues to grow, the need for innovative solutions to manage radio resources efficiently becomes more critical. This chapter explores the integration of blockchain technology to enhance the security and efficiency of radio resource management (RRM) in fifth generation (5G) and beyond networks.

1.1 Overview of 5G and beyond networks

The 5G of mobile networks represents a significant leap in wireless communication technology, promising unprecedented speed, ultralow latency

DOI: 10.1201/9781003514336-

and massive connectivity. These capabilities are essential for supporting a wide array of applications, from autonomous vehicles and smart cities to enhanced mobile broadband and the Internet of Things (IoT) [1]. As we move beyond 5G, the focus continues to be on improving these parameters to support even more sophisticated and diverse use cases, highlighting the need for efficient and secure RRM.

1.2 Importance of Radio Resource Management

RRM is crucial in 5G and beyond networks, as it involves the allocation and optimization of radio frequencies to ensure seamless connectivity and optimal network performance. Effective RRM minimizes interference, maximizes spectral efficiency, and guarantees quality of service (QoS) for diverse applications [2]. Given the complexity and dynamic nature of modern wireless networks, traditional RRM methods face significant challenges in terms of scalability, latency, and security.

1.3 Introduction to blockchain technology

Blockchain technology, initially conceptualized for digital currencies like Bitcoin, has emerged as a transformative solution across various industries due to its decentralized, transparent, and secure nature. It consists of a distributed ledger that records transactions in a verifiable and immutable manner [3]. Blockchain's potential to enhance security, transparency, and efficiency makes it an attractive solution for addressing the inherent challenges in traditional RRM.

1.4 Objectives of the chapter

This chapter aims to explore the integration of blockchain technology into RRM for 5G and beyond networks. It will provide a comprehensive overview of blockchain fundamentals, its current applications in telecommunications, and the specific challenges faced by traditional RRM. This chapter will also propose an architectural framework for blockchain-based RRM, highlight its potential to enhance efficiency and security, and present real-world case studies and future research directions.

2 BASICS OF BLOCKCHAIN TECHNOLOGY

Blockchain technology, a revolutionary concept initially devised for digital currencies, has since expanded its utility across numerous industries due to its decentralized and secure nature. Understanding the fundamental aspects of blockchain is essential to appreciating its potential applications in telecommunication networks, particularly in enhancing RRM.

2.1 History and evolution

Blockchain technology, which underpins digital currencies like Bitcoin, was first introduced by an individual or group known as Satoshi Nakamoto in 2008. The technology was designed to enable peer-to-peer transactions without the need for a trusted third party. The initial success of Bitcoin sparked interest in the underlying technology, leading to the development of various blockchain platforms and applications beyond digital currencies [3,4]. Over time, blockchain has evolved from a niche technology into a foundational framework for developing decentralized applications across multiple sectors.

2.1.1 Key components: blocks, chains, and nodes

At its core, a blockchain is a decentralized ledger composed of a sequence of blocks, each containing a list of transactions. Each block is cryptographically linked to the previous one, forming a chain. This linkage ensures that any alteration in a block would invalidate subsequent blocks, thereby providing immutability. Nodes are individual computers that maintain copies of the blockchain and participate in the network by validating and propagating transactions [5].

2.2 Consensus mechanisms: proof of work, proof of stake, and others

Consensus mechanisms are protocols used to achieve agreement on the blockchain's state among distributed nodes. The most well-known is proof of work (PoW), used by Bitcoin, where nodes (miners) solve complex mathematical problems to validate transactions and create new blocks. Another popular mechanism is proof of stake (PoS), where validators are chosen based on the number of tokens they hold and are willing to "stake" as collateral. Other mechanisms include delegated proof of stake (DPoS), Byzantine fault tolerance (BFT), and more, each offering different trade-offs in terms of security, efficiency, and decentralization [6].

2.3 Smart contracts

Smart contracts are self-executing contracts with the terms of the agreement directly written into code. These contracts automatically execute and enforce the terms when predefined conditions are met. Smart contracts run on blockchain platforms such as Ethereum, thereby allowing for decentralized and trustless applications across various fields, from finance to supply chain management [7].

3 BLOCKCHAIN IN TELECOMMUNICATION NETWORKS

The telecommunications industry, with its complex and extensive infrastructure, is constantly seeking innovative solutions to enhance efficiency, security, and transparency. Blockchain technology, with its decentralized and immutable nature, offers significant potential to address many of the challenges faced by telecom operators. By integrating blockchain, telecom companies can streamline operations, reduce fraud, and improve the overall customer experience. Blockchain technology has begun to make significant inroads into the telecommunications sector, offering a range of applications that enhance security, efficiency, and transparency. Certain applications are listed as following:

- **Fraud prevention and identity management**: One of the primary applications of blockchain in telecom is fraud prevention. With blockchain's immutable ledger, telecommunications companies can create secure and verifiable records of transactions and customer identities. This application helps in combating identity theft, SIM card fraud, and roaming fraud. For instance, Telefónica has implemented blockchain solutions to verify user identities, reducing fraudulent activities significantly [2].
- **Secure transactions and billing**: Blockchain technology facilitates secure and transparent financial transactions, making it ideal for managing billing and payment systems in telecom. Smart contracts can automate billing processes and enforce payment terms, reducing disputes and delays. Companies like Deutsche Telekom are exploring blockchain-based billing systems to ensure accuracy and transparency in transactions [8].
- **Inter-carrier settlements**: Inter-carrier settlements, which involve the complex process of revenue sharing between different telecom operators, can be streamlined using blockchain. By creating a shared, transparent ledger, blockchain reduces the time and cost associated with these settlements. For example, the Carrier Blockchain Study Group (CBSG), a consortium of global telecom carriers, is developing blockchain solutions for faster and more efficient inter-carrier settlements [9].

- **Supply chain management**: Blockchain enhances supply chain management by providing an immutable record of each transaction and component within the supply chain. This transparency ensures accountability and reduces the risk of counterfeit products. Telecom companies like Vodafone are using blockchain to track and manage their supply chain more effectively, ensuring the authenticity of network equipment [10].

3.1 Potential benefits for 5G networks

As 5G networks promise to deliver ultrareliable, low-latency communication, integrating blockchain technology can provide several key benefits:

- **Enhanced security**: 5G networks involve numerous connected devices, each potentially vulnerable to cyberattacks. Blockchain's decentralized and immutable nature can secure these devices by ensuring data integrity and preventing unauthorized access. For instance, blockchain can secure IoT devices connected through 5G by validating and recording each transaction in an unalterable ledger, thereby reducing the risk of cyber threats [11].
- **Efficient spectrum management**: Managing radio frequencies efficiently is crucial for 5G networks. Blockchain can provide a transparent and decentralized mechanism for spectrum allocation, ensuring fair and efficient use of this valuable resource. Smart contracts can automate spectrum auctions and allocations, reducing administrative overhead and enhancing operational efficiency [12].
- **Decentralized network management**: Blockchain enables decentralized network management, thereby reducing reliance on central authorities. This decentralization enhances the resilience and flexibility of 5G networks. For example, a blockchain-based system can enable decentralized control of network functions, leading to more efficient and responsive network management [13].
- **Automated service-level agreements (SLAs)**: Smart contracts can automate SLAs between service providers and customers, ensuring that the terms are met without manual intervention. This automation increases trust and reduces the need for intermediaries, leading to faster and more reliable service delivery [14].

3.2 Case studies and industry adoption

- **Telefónica's blockchain projects**: Telefónica, a leading global telecom provider, has launched multiple blockchain initiatives. One notable project is TrustOS, a blockchain-based platform designed to provide digital trust services. TrustOS enhances data integrity, transparency, and security across various applications, including identity verification and supply chain management [15].
- **Deutsche Telekom and T-Labs**: Deutsche Telekom's research and development arm, T-Labs, is actively exploring blockchain applications. One significant initiative is the development of blockchain-based solutions for secure and transparent inter-carrier settlements. This project aims to simplify the settlement process, reduce fraud, and enhance trust between carriers [8].
- **Carrier Blockchain Study Group (CBSG)**: The CBSG, a consortium of global telecom carriers, is pioneering the use of blockchain for various applications, including inter-carrier payments, digital identity verification, and mobile wallet services. By leveraging blockchain, CBSG aims to create a more efficient and secure ecosystem for telecommunications services [9].
- **Vodafone's supply chain management**: Vodafone is using blockchain to enhance its supply chain operations. By tracking each component from manufacturing to deployment, Vodafone ensures the authenticity and integrity of its network equipment. This blockchain implementation has led to improved transparency, reduced fraud, and enhanced operational efficiency [10].

4 CHALLENGES IN TRADITIONAL RADIO RESOURCE MANAGEMENT

As the demand for wireless communication grows with the advent of 5G and beyond networks, traditional RRM faces numerous challenges. These challenges span from security vulnerabilities to inefficiencies, scalability concerns, and issues related to trust and transparency. Addressing these challenges is crucial for optimizing the performance and reliability of modern telecommunications networks.

- Traditional RRM systems in telecommunications face significant security vulnerabilities, primarily due to their centralized architecture. Cyberattacks and data breaches are major concerns, as a successful attack on the central authority can compromise the entire network, leading to the loss of sensitive information and disruption of services. The lack of robust end-to-end encryption in traditional RRM systems leaves data transmitted over the network susceptible to interception and eavesdropping, posing a serious risk for applications that require high levels of security, such as financial transactions and healthcare communications [16,17]. Furthermore, traditional RRM systems often struggle to prevent unauthorized access effectively. As the number of devices and users in the network increases, ensuring that only authorized entities access network resources becomes increasingly challenging, leading to unauthorized spectrum usage and potential interference with legitimate communications [18].
- Traditional RRM systems often result in significant inefficiencies and latency issues, which can severely impact network performance. One of the primary inefficiencies arises from suboptimal spectrum utilization due to static allocation methods. These methods fail to adapt dynamically to changing network conditions, leading to inefficient use of available resources, with some frequency bands being underutilized while others experience congestion [19]. The centralized nature of traditional RRM also introduces significant latency in resource allocation and management. Decisions made by a central authority take time to propagate through the network, causing delays in responding to real-time changes in demand. This latency is particularly detrimental to applications requiring low-latency communication, such as autonomous vehicles and remote surgery [20]. Additionally, managing and coordinating resources across a complex and heterogeneous network requires substantial administrative effort. Traditional RRM systems often rely on manual processes and rigid policies, increasing the administrative burden and reducing operational efficiency [21].
- As the telecommunications industry transitions to 5G and beyond networks, scalability becomes a major concern for traditional RRM systems. These systems struggle to handle the massive connectivity demands expected in 5G networks, which support a significantly higher number of connected devices than previous generations.

The scalability challenge is exacerbated by the increased volume of data traffic, which traditional systems find difficult to manage efficiently while maintaining performance [22]. Interference management is another critical scalability issue. With the increasing density of devices in the network, traditional RRM techniques are often inadequate for managing interference in highly dense environments, leading to degraded performance and reduced QoS [23]. Furthermore, 5G networks are characterized by their heterogeneity, encompassing various types of cells, frequencies, and technologies. Traditional RRM systems are not well-equipped to handle the complexities of heterogeneous networks, resulting in inefficient resource allocation and potential conflicts between different network segments [24].

- Traditional RRM systems face significant trust and transparency issues, primarily due to their centralized control structure. This centralization creates single points of failure, which can compromise the trustworthiness of the entire network. If the central authority is compromised or fails, the entire system can be disrupted, undermining the network's reliability [13]. Moreover, the lack of transparency in how resources are allocated and managed by traditional RRM systems can lead to disputes and mistrust among network users and operators. Without a clear, verifiable record of resource allocation decisions, stakeholders have no way to ensure that resources are being used fairly and efficiently [25]. Enforcing SLAs effectively is another challenge. SLAs are critical for defining the QoS guarantees between service providers and customers, but traditional RRM systems often lack the automated and transparent mechanisms needed to enforce these agreements reliably, leading to unmet performance guarantees and potential disputes between parties [26]

5 INTEGRATING BLOCKCHAIN FOR RRM

The integration of blockchain technology into RRM offers promising solutions to address the challenges faced by traditional RRM systems. This section delves into the architectural framework for blockchain-based RRM, explores different deployment models, discusses interoperability with existing systems, and explains how blockchain can enhance secure spectrum sharing and allocation.

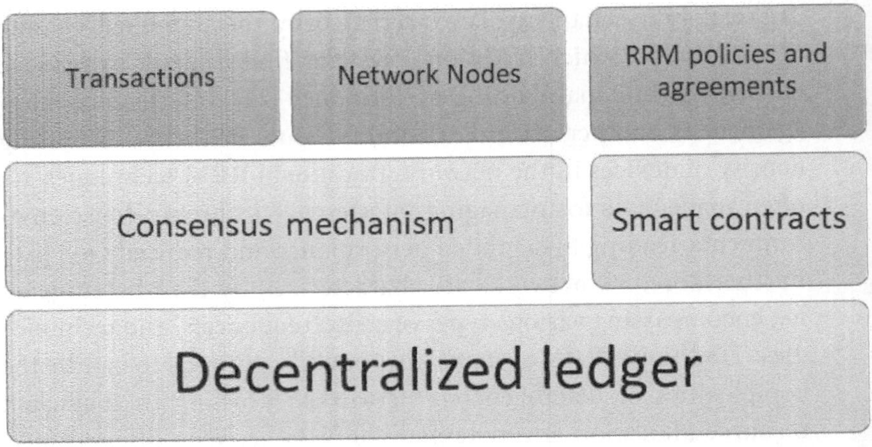

FIGURE 4.1 Architecture of blockchain-based 5G RRM.

5.1 Architectural framework for blockchain-based RRM

A blockchain-based RRM system shown in Figure 4.1 comprises several key components: decentralized ledger, consensus mechanism, smart contracts, and network nodes.

- The decentralized ledger serves as the backbone, recording all transactions and resource allocations in an immutable manner. Each network node, representing a telecom entity or a base station, maintains a copy of the ledger, ensuring transparency and redundancy.
- The consensus mechanism is crucial for validating transactions and ensuring that all nodes agree on the state of the blockchain. Popular mechanisms like PoW and PoS can be adapted for telecom networks. For instance, PoS can be leveraged to prioritize transactions based on the stake held by different network entities, thereby ensuring efficient resource management.
- Smart contracts automate the execution of RRM policies and agreements. They can be programmed to manage spectrum allocation, enforce SLAs, and handle inter-operator settlements. These contracts run on the blockchain, providing a trustless environment where agreements are automatically executed when predefined conditions are met. This reduces the need for intermediaries and minimizes administrative overhead.

In a blockchain-based RRM architecture, interoperability with existing telecom infrastructure is critical. Gateways and application programming interfaces (APIs) can facilitate communication between blockchain nodes and traditional network components, ensuring seamless integration. Additionally, distributed databases and decentralized storage solutions can enhance data management and retrieval efficiency.

5.2 Deployment models: public, private, and consortium blockchains

Blockchain deployment models significantly impact the functionality and performance of RRM systems. The three primary models are public, private, and consortium blockchains.

Public blockchains: Public blockchains, such as Bitcoin and Ethereum, are open to anyone and are characterized by high transparency and decentralization. In the context of RRM, public blockchains can provide a transparent and secure environment for spectrum auctions and public resource allocations. However, they may face scalability issues and higher latency due to the large number of participants and the complexity of consensus mechanisms.

Private blockchains: Private blockchains restrict access to a single organization or a group of trusted entities. This model offers enhanced control over the network and better performance in terms of transaction speed and scalability. For telecom operators, private blockchains can be used for internal RRM processes, such as intra-operator resource management and private spectrum allocation. The main challenge lies in maintaining trust and security within a smaller, centralized framework.

Consortium blockchains: Consortium blockchains are semi-decentralized, with control shared among a group of organizations. This model strikes a balance between the transparency of public blockchains and the control of private blockchains. In telecom, consortium blockchains can facilitate inter-operator collaborations, such as spectrum sharing and roaming agreements. By involving multiple stakeholders, consortium blockchains enhance trust and reduce the risk of single points of failure, while maintaining relatively high performance and scalability.

5.3 Interoperability with existing systems

For blockchain-based RRM to be effective, it must seamlessly integrate with existing telecom systems. This requires developing interoperability solutions that enable communication and data exchange between blockchain platforms and traditional network components.

Gateways and APIs: Gateways act as intermediaries that translate data and commands between blockchain nodes and legacy systems. APIs provide standardized interfaces for different systems to interact with the blockchain. These tools are essential for ensuring that existing telecom infrastructure can communicate with the blockchain without significant modifications.

Hybrid systems: A hybrid approach can combine blockchain technology with traditional centralized systems. For example, blockchain can be used for critical functions like secure spectrum allocation, while conventional systems handle routine network management tasks. This hybrid model allows for a gradual transition to blockchain-based RRM, minimizing disruption and leveraging the strengths of both technologies.

Standardization efforts: Industry standards for blockchain interoperability are essential for ensuring compatibility across different platforms and networks. Organizations like the IEEE and the International Telecommunication Union (ITU) are working on developing standards that define protocols, data formats, and security measures for blockchain integration in telecom. Adopting these standards can facilitate widespread adoption and interoperability of blockchain-based RRM systems.

6 BLOCKCHAIN FOR SECURE SPECTRUM SHARING AND ALLOCATION

Spectrum sharing and allocation are critical aspects of RRM that can greatly benefit from blockchain technology. Blockchain's inherent properties of transparency, immutability, and decentralization make it an ideal solution for managing spectrum resources securely and efficiently.

Blockchain can facilitate *decentralized and transparent spectrum auctions*, where bids and transactions are recorded on the blockchain. This ensures that all participants have equal access to information and that the auction process is fair and tamper-proof. Smart contracts can automate the bidding process, execute winning bids, and allocate spectrum resources without human intervention, reducing the risk of fraud and manipulation [20].

Traditional spectrum allocation methods are often static and inefficient. Blockchain enables *dynamic spectrum access* by allowing real-time monitoring and reallocation of spectrum resources. Smart contracts can enforce policies that dynamically assign spectrum based on current demand and usage patterns, optimizing resource utilization and reducing interference [13].

By recording all spectrum transactions on a decentralized ledger, blockchain enhances the *security and trustworthiness of the spectrum management process*. Each transaction is cryptographically secured, ensuring data integrity and preventing unauthorized modifications. This transparency builds trust among operators and regulators, promoting more efficient and cooperative spectrum management [27].

Blockchain facilitates secure and efficient spectrum sharing between different operators. In a consortium blockchain model, operators can *collaborate on spectrum usage*, dynamically allocating resources based on agreements encoded in smart contracts. This collaborative approach can reduce spectrum wastage and improve overall network performance [28].

7 ENHANCING EFFICIENCY WITH BLOCKCHAIN

Integrating blockchain technology into RRM systems can significantly enhance efficiency by reducing latency in resource allocation, automating processes through smart contracts, and enabling decentralized decision-making. This section explores how blockchain contributes to these improvements and provides examples and use cases.

- **Reducing latency in resource allocation**: Traditional RRM systems often rely on centralized decision-making, which can introduce significant delays due to the time required to process and propagate information through the network. Blockchain, with its decentralized nature, allows for faster and more efficient resource allocation. In a blockchain-based RRM system, the *distributed ledger* ensures that all network nodes have real-time access to the same information. This eliminates the need for a central authority to validate and distribute updates, significantly reducing the time required to allocate resources. Each node can independently verify transactions and make decisions based on the most up-to-date information available [11]. Smart contracts can *automate the resource allocation* process by executing predefined rules without the need for human intervention. For example, a smart contract could be programmed to automatically allocate additional spectrum to a base station when traffic levels exceed a certain threshold. This automation reduces latency by enabling immediate responses to changing network conditions, ensuring optimal resource utilization at all times [29].

- **Automated smart contracts for resource management**: Smart contracts are self-executing contracts with the terms of the agreement directly written into code. They run on the blockchain and automatically enforce the conditions specified in the contract. This feature is particularly useful for automating resource management tasks in telecom networks. Smart contracts can *dynamically allocate spectrum resources* based on real-time network demands. For example, during peak hours, a smart contract could redistribute spectrum from underutilized areas to high-demand zones. This dynamic allocation ensures that spectrum resources are used efficiently, improving overall network performance [20]. *Billing and settlements* between operators can be automated using smart contracts. These contracts can track resource usage and automatically generate invoices based on predefined rates. Payments can be processed instantly, reducing administrative overhead and ensuring timely settlements. This automation not only improves efficiency but also enhances transparency and reduces the risk of disputes [8]. Smart contracts can enforce SLAs by continuously monitoring network performance and automatically applying penalties or rewards based on compliance. For instance, if an operator fails to meet the agreed-upon QoS, a smart contract can automatically trigger a compensation payment to the affected customers. This ensures that SLAs are upheld without manual intervention, improving customer satisfaction and trust [29].
- **Decentralized decision-making processes**: Decentralized decision-making is a fundamental characteristic of blockchain technology that enhances the efficiency and resilience of RRM systems. By distributing decision-making authority across multiple nodes, blockchain reduces the reliance on a central authority and enables faster, more robust responses to network changes. *Consensus mechanisms*, such as PoS or DPoS, allow network participants to agree on the state of the blockchain. These mechanisms enable decentralized decision-making by ensuring that all nodes reach a consensus on resource allocations and other critical decisions. This process is faster and more secure than traditional centralized methods, which are vulnerable to single points of failure [6]. Blockchain enables

collaborative resource management by allowing multiple operators to participate in decision-making processes. For example, in a consortium blockchain model, operators can collectively decide on spectrum sharing arrangements and other resource allocations. This collaboration reduces conflicts and ensures that resources are allocated in a way that benefits all parties involved [13]. Decentralized decision-making enhances the *resilience of RRM systems* by eliminating single points of failure. If one node fails or is compromised, the remaining nodes can continue to operate and maintain the network. This resilience is particularly important in scenarios where network reliability is critical, such as emergency communications and public safety networks [27].

8 COMPARATIVE ANALYSIS WITH TRADITIONAL METHODS

TABLE 4.1 Traditional RRM vs. blockchain-based RRM

Aspect	Traditional RRM	Blockchain-based RRM
Security and trust	Centralized systems with single points of failure and higher susceptibility to cyberattacks.	Decentralized, immutable ledger enhances security and reduces single points of failure. Increased transparency builds trust [3].
Efficiency and cost	Manual processes and intermediaries increase administrative overhead and costs. Transactions and settlements are time-consuming.	Smart contracts automate processes, reducing administrative overhead and costs. Faster transactions and settlements without intermediaries [8].
Scalability and flexibility	Struggles with scalability and dynamic resource allocation, especially with the increasing number of connected devices and heterogeneous networks.	More scalable and flexible, capable of dynamically allocating resources and managing complex interactions in a decentralized manner [20].
Latency in resource allocation	Centralized decision-making introduces significant latency, causing delays in responding to real-time network changes.	Decentralized nature and automated smart contracts significantly reduce latency, allowing immediate responses to network conditions.

(*Continued*)

TABLE 4.1 (*Continued*) Traditional RRM vs. blockchain-based RRM

Aspect	Traditional RRM	Blockchain-based RRM
Fraud prevention and transparency	Vulnerable to fraud due to lack of transparency and centralized control. Fraud detection relies on manual oversight.	Transparent and verifiable transactions reduce fraud. Smart contracts automate fraud detection and enforcement of agreements.
Interoperability	Challenges in integrating with new technologies and systems, requiring significant modifications and standardization efforts.	Easier integration with emerging technologies like AI and IoT. Standardized protocols and APIs facilitate interoperability with existing and new systems.
Privacy and data protection	Limited privacy protection mechanisms, with centralized databases at risk of breaches and unauthorized access.	Enhanced privacy through pseudonymous transactions and selective disclosure mechanisms. Advanced cryptographic techniques protect data integrity.

9 FUTURE DIRECTIONS AND RESEARCH OPPORTUNITIES

As blockchain technology continues to evolve, its integration with 5G and beyond networks presents numerous opportunities for innovation and research. The potential for blockchain to enhance security, efficiency, and transparency in RRM is vast, yet several challenges and areas for further exploration remain. One of the most promising areas for future research is the integration of blockchain with emerging technologies such as artificial intelligence (AI) and the IoT. AI can be leveraged to optimize blockchain operations, such as enhancing consensus algorithms and improving the efficiency of smart contracts. Machine learning models can predict network conditions and dynamically adjust blockchain-based RRM processes, leading to more efficient resource allocation and management. The combination of AI and blockchain can create intelligent, autonomous systems that enhance the performance and reliability of telecom networks. The integration of blockchain with IoT in 5G networks is another critical area for research. IoT devices generate massive amounts of data, and managing this data securely and efficiently is a significant challenge. Blockchain can provide a decentralized and immutable ledger for IoT data, ensuring its integrity and enhancing security. Research can focus on developing lightweight blockchain protocols suitable for resource-constrained IoT devices and exploring new consensus mechanisms that are both efficient and secure for IoT environments.

Interoperability between different blockchain platforms and with existing telecom infrastructure is another important research area. As multiple blockchain platforms emerge, ensuring they can interoperate seamlessly is crucial for widespread adoption. Developing standardized protocols and interfaces that enable different blockchains to communicate and share data will be key to achieving this interoperability. Additionally, research can explore how to integrate blockchain with legacy telecom systems without causing significant disruption, ensuring a smooth transition to blockchain-based RRM. Scalability remains a major challenge for blockchain technology, especially in the context of 5G networks, which involve a vast number of connected devices and high data throughput. Research can focus on developing scalable blockchain solutions that can handle the high transaction volumes and low latency requirements of 5G. Layer 2 scaling solutions, such as off-chain transactions and sidechains, offer promising approaches to improving blockchain scalability and performance.

Privacy and data protection are also critical areas for research. While blockchain provides transparency and security, it can also expose sensitive information if not properly managed. Research can explore advanced cryptographic techniques, such as zero-knowledge proofs and homomorphic encryption, to enhance privacy on blockchain networks. These techniques can enable the validation of transactions without revealing sensitive information, ensuring user privacy while maintaining the security and integrity of the blockchain. The economic and regulatory implications of blockchain in telecom networks also warrant further investigation. The deployment of blockchain can disrupt traditional business models and regulatory frameworks, raising questions about governance, compliance, and the role of regulatory bodies. Research can examine the economic impact of blockchain adoption, develop new business models that leverage blockchain's capabilities, and propose regulatory frameworks that ensure fair competition and protect consumer interests.

Finally, real-world implementations and pilot projects are essential for testing and validating blockchain applications in telecom networks. Collaborative efforts between academia, industry, and regulatory bodies can facilitate these pilot projects, providing valuable insights and driving the adoption of blockchain technology. Case studies of successful implementations can serve as blueprints for future projects, highlighting best practices and lessons learned. In summary, the integration of blockchain with 5G and beyond networks presents exciting opportunities for innovation and research. By exploring the intersection of blockchain with AI,

IoT, and other emerging technologies, addressing challenges related to scalability, interoperability, and privacy, and examining the economic and regulatory implications, researchers can unlock the full potential of blockchain for enhancing RRM in modern telecom networks. The continuous evolution of blockchain technology promises to transform the telecommunications industry, paving the way for more secure, efficient, and transparent network management.

10 CONCLUSION

Blockchain technology holds significant promise for transforming RRM in 5G and beyond networks. By leveraging blockchain's decentralized, immutable, and transparent nature, telecom operators can address many of the challenges faced by traditional RRM systems, including security vulnerabilities, inefficiencies, scalability issues, and trust concerns. Blockchain enhances security through its cryptographic techniques, decentralized authentication, and immutable ledger, preventing fraud and unauthorized access while ensuring data integrity. The use of smart contracts automates resource management processes, reducing latency and administrative overhead. Furthermore, blockchain enables decentralized decision-making, improving the resilience and flexibility of telecom networks.

Real-world implementations, such as those by Telefónica, Deutsche Telekom, and the CBSG, demonstrate the practical benefits of blockchain in telecom, including improved identity management, efficient interoperator settlements, and enhanced supply chain management. These case studies highlight the potential for blockchain to revolutionize the telecommunications industry. Looking ahead, integrating blockchain with emerging technologies like AI and IoT, developing scalable and interoperable solutions, and addressing privacy and regulatory challenges will be critical for realizing the full potential of blockchain in telecom. By continuing to explore these areas through research and pilot projects, the telecom industry can harness blockchain to create more secure, efficient, and transparent networks for the future.

REFERENCES

1. Andrews, J. G., Buzzi, S., Choi, W., Hanly, S. V., Lozano, A., Soong, A. C., & Zhang, J. C. (2014). What will 5G be? *IEEE Journal on Selected Areas in Communications*, 32(6), 1065–1082.
2. Wang, C. X., Haider, F., Gao, X., You, X. H., Yang, Y., Yuan, D., ... & Aggoune, H. M. (2018). Cellular architecture and key technologies for 5G wireless communication networks. *IEEE Communications Magazine*, 52(2), 122–130.

3. Nakamoto, S. (2008). *Bitcoin: A Peer-to-Peer Electronic Cash System.* Available at SSRN: https://ssrn.com/abstract=3440802 or http://dx.doi.org/10.2139/ssrn.3440802
4. Swan, M. (2015). *Blockchain: Blueprint for a New Economy.* O'Reilly Media, Inc.
5. Wood, G. (2014). Ethereum: A secure decentralised generalised transaction ledger. *Ethereum Project Yellow Paper,* 151, 1–32.
6. Buterin, V. (2013). Ethereum white paper. *A Next Generation Smart Contract & Decentralized Application Platform. Ethereum White Paper, @ in proceedings,* 1-36.
7. Szabo, N. (1997). *The Idea of Smart Contracts. Nick Szabo's Papers and Concise Tutorials.*
8. Mendling, J., Weber, I., Van Der Aalst, W. M., Vom Brocke, J., Cabanillas, C., Daniel, F., ... & Zhu, L. (2018). Blockchains for business process management – Challenges and opportunities. *ACM Transactions on Management Information Systems (TMIS),* 9(1), 1–16.
9. Deng, W., Huang, T., & Wang, H. (2022). A review of the key technology in a blockchain building decentralized trust platform. *Mathematics,* 11(1), 101.
10. Tortorella, G., Fogliatto, F. S., Gao, S., & Chan, T. K. (2022). Contributions of Industry 4.0 to supply chain resilience. *The International Journal of Logistics Management,* 33(2), 547-566.
11. Dorri, A., Kanhere, S. S., Jurdak, R., & Gauravaram, P. (2017). Blockchain for IoT security and privacy: The case study of a smart home. *2017 IEEE International Conference on Pervasive Computing and Communications Workshops (PerCom Workshops),* 618–623.
12. Zhang, P., White, J., Schmidt, D.C., & Lenz, G. (2017). *Applying Software Patterns to Address Interoperability in Blockchain-based Healthcare Apps.* ArXiv, abs/1706.03700.
13. Guvenc, I., Arslan, H., Chong, P. H. J., & Poor, H. V. (2017). Guest editorial 5G for mission critical IoT communications. *IEEE Network,* 31(6), 22–23.
14. Yu, R., Xie, S., Zhang, Y., & Xiao, W. (2015). Security and privacy in 5G networks: Challenges and solutions. *IEEE Communications Magazine,* 53(12), 12–19.
15. Li, X., Lu, R., Liang, X., Shen, X., Chen, J., & Lin, X. (2018). Smart community: An internet of Things application. *IEEE Communications Magazine,* 55(2), 68–75.
16. Cheng, X., Zhang, R., & Sun, L. (2018). Spectrum sharing for 5G/IoT: A survey. *IEEE Internet of Things Journal,* 6(2), 1168–1193.
17. Ning, Z., Xia, F., Ullah, N., Kong, X., Hu, X., & Al-Dhelaan, A. (2019). Vehicular social networks: Enabling smart mobility. *IEEE Communications Magazine,* 55(5), 49–55.
18. Rupasinghe, N., Yazar, A., Karaca, O., Ozdemir, S., & Arslan, H. (2018). Spectrum management in cognitive radio networks with heterogeneity. *IEEE Communications Magazine,* 56(4), 182–189.
19. Bangerter, B., Talwar, S., Arefi, R., & Stewart, K. (2014). Networks and devices for the 5G era. *IEEE Communications Magazine,* 52(2), 90–96.

20. Xu, X., Weber, I., Staples, M., Zhu, L., Bosch, J., Bass, L., ... & Rimba, P. (2016). A taxonomy of blockchain-based systems for architecture design. *Proceedings of the 2016 International Conference on Software Architecture*, Gothenburg, Sweden, 243–252.
21. Mijumbi, R., Serrat, J., Gorricho, J. L., Bouten, N., De Turck, F., & Davy, S. (2016). Network function virtualization: State-of-the-art and research challenges. *IEEE Communications Surveys & Tutorials*, 18(1), 236–262.
22. Kim, J. S., & Lee, Y. G. (2019). Blockchain-based authentication and authorization for IoT devices in 5G network. *Journal of Network and Computer Applications*, 123, 32–40.
23. Reyna, A., Martín, C., Chen, J., Soler, E., & Díaz, M. (2018). On blockchain and its integration with IoT. Challenges and opportunities. *Future Generation Computer Systems*, 88, 173–190.
24. Telefónica. (2020). TrustOS: Enhancing digital trust in the new digital era. [Online] Available at: https://www.telefonica.com/es/wp-content/uploads/sites/4/2006/01/hr030106.pdf
25. Lemieux, V. L. (2016). Trusting records: Is blockchain technology the answer? *Records Management Journal*, 26(2), 110–139.
26. Meiklejohn, S., Pomarole, M., Jordan, G., Levchenko, K., McCoy, D., Voelker, G. M., & Savage, S. (2013). A fistful of bitcoins: Characterizing payments among men with no names. *Proceedings of the 2013 Conference on Internet Measurement Conference*, Barcelona, Spain, 127–140.
27. Zyskind, G., Nathan, O., & Pentland, A. (2015). Decentralizing privacy: Using blockchain to protect personal data. *2015 IEEE Security and Privacy Workshops*, 180–184.
28. Hawig, D., Zhou, C., Fuhrhop, S., Fialho, A. S., & Ramachandran, N. (2019). Designing a distributed ledger technology system for interoperable and general data protection regulation-compliant health data exchange: A use case in blood glucose data. *Journal of Medical Internet Research*, 21(6), e13665.
29. Tachibana, R., Shioji, H., & Hirose, Y. (2018). Spectrum sharing by blockchains and smart contracts. *2018 IEEE International Symposium on Dynamic Spectrum Access Networks (DySPAN)*, Seoul, South Korea, 1–6.

CHAPTER 5

Federated learning for intelligent network management in 5G

B. Sundaravadivazhagan, N.A. Natraj, and S. Gopinath

1 INTRODUCTION

5G wireless technologies can transform communication networks. Its quicker speeds, less delay, and other benefits will power a variety of present and future technology. The 5G network has faster connections, higher capacity, and lower latency (less than 1 millisecond vs. 30 and 100 for 4G and 3G). Service providers claim theoretical speeds of up to 10 Gps, 100 times faster than 4G networks' top throughput of 100 Mbps. While speeds may vary and are unlikely to reach the maximum, the technology will offer a better experience and more possibilities than current wireless technologies. Numerous technology businesses and suppliers are developing and deploying these networks in experimental or commercial settings. Mobile broadband is leading the way in high-speed data transfers for mobile devices and other applications.

1.1 5G technologies and methodologies

In anticipation of the integration of 5G standards, numerous technologies and methodologies have been developed. 5G will be capable of offering a

DOI: 10.1201/9781003514336-5

dynamic and adaptable service as a result of these novel technologies and approach. Among the technologies that are currently being developed for 5G are:

a. Millimeter-wave communications

   These frequencies are significantly higher in the frequency range. It enables the utilization of a wide channel bandwidth of up to 2 GHz and a variety of new spectrums [1]. Nevertheless, this introduces supplementary challenges to handset development, as the utmost frequencies and bandwidths are typically between 10 and 20 MHz. In terms of circuit design, technology, and system utilization, frequencies exceeding 50 GHz pose challenging obstacles for 5G, as they do not travel over a significant distance. They are entirely consumed by obstacles. The allocation of spectrum to 5G varies among different countries.

b. Waveforms

   The introduction of novel waveforms has generated considerable curiosity. OFDM is successfully implemented in numerous high-data-rate systems, including 4G LTE. Nevertheless, it is constrained in a few specific situations. GFDM, universal filtered multicarrier, filter bank multi-carrier, UFMC, generalized frequency division multiplexing, and FBMC are the other waveform formats that are discussed. A waveform that is flawless does not exist. OFDMA is employed because it provides exceptional overall performance without relying excessively on processing capacity [2].

c. Multiple accesses

   Numerous novel access schemes are currently being investigated for 5G technology. OFDMA, IDMA, SCMA, NOMA, MUSA, and PDMA comprise the techniques that are currently being evaluated. Nevertheless, OFDMA is the most likely format [2].

d. Massive MIMO with beam guidance

   Despite the fact that MIMO is employed in a variety of applications, including Wi-Fi and LTE, the antenna designs are relatively restricted. Microwave frequencies enable the application of multiple antennas on a single piece of equipment as a result of the antenna diameters and spacing in wavelength terms. This configuration would enable the piloting of beams to provide superior performance.

e. Networks that are dense

The reduction of cell size results in a more efficient and effective utilization of the available spectrum. Techniques must be implemented to guarantee that the micro-cells that are present within the macro-network and subsequently deployed as femtocells function as intended. The addition of additional cells to a network presents significant obstacles, and strategies are being developed to prevent this from occurring.

1.2 Utilization and applications of 5G

Allowing new kinds of applications is the aim of 5G technology. It is not simply a progression of the prior cellular generation, but also a ground-breaking technology that is supposed to remove access, capacity, performance, and latency barriers on worldwide connectivity.

a. Enhanced mobile broadband

The objective of enhanced mobile broadband (eMBB) is to enable revolutionary mobile applications like immersive augmented reality (AR) and virtual reality (VR) on the go, as well as high-definition video streaming. When it comes to 5G, one would expect rapid speeds. However, mobility and capability may not always be required. A 5G network, for example, could be heavily used in a certain place. In such a circumstance, mobility requirements would be minor, but capacity requirements would be extremely high. To maintain that high-speed experience, however, movement in a moving vehicle—particularly an airplane or train—would necessitate enormous mobility. Bigger data amounts and an improved user experience are made possible by the rise of eMBB, which reflects that of modern mobile broadband. Supporting increased end-user data rates is a perfect example. Typical eMBB applications include 4K and 8K Ultra HD video in three dimensions, kiosks for broadband, cloud gaming, and tactile internet, holographic VR and AR, remote classroom, simulated and trained in real time.

b. Massive machine type communication

It is beneficial to consider 5G massive machine type communication (mMTC) as a means of connecting a significant number of items in the Internet of Things (IoT). The mMTC concept is conceived to connect a vast number of devices in a specific area—up to 1 million devices per square kilometer—with minimal energy consumption and data rate requirements. mMTC has the potential to be used in smart agriculture, where sensors can monitor and respond to minor changes in growing conditions at locations across a large area [3]. This would enable targeted, near-real-time responses to optimize growth rates. It could also facilitate similar functions, such as near-real-time data collection and response, in urban areas to assist in the management of traffic or a more intelligent electric grid; in factories to assist in the management of production, quality assurance, or shipping; in shipping fleets to monitor routes, efficiency, and maintenance requirements; and in other applications. It also describes services including actuators, equipment monitoring devices, and remote sensors with a big device population. For such services, two prerequisites are low device costs and effective energy consumption from devices. This makes the gadgets' batteries last for several years. Just a certain quantity of data is generated and received by each device. High data rate support has so lost some of its significance in this situation.

c. Ultra-reliable low-latency communication

Ultra-reliable low-latency communication (URLLC) is an important part of 5G because it makes low-latency and reliable transmission possible for mission-critical uses like VR and AR, self-driving cars, and automating factories. It promises a high level of dependability, very low delay (1 ms), and a certain quality of service (QoS) [1]. URLLC services are used where we require a very low latency and a very high dependability. Among the examples are automatic control, factory automation, and traffic safety. Overall, 5G will transform cell phone use in 1 Gbps or higher bandwidth places. When 5G is used with VoIP, call volume and data transmission will be unparalleled. 5G will enable ubiquitous networks, radio resource management, HAPS systems, and more.

1.3 Importance of efficient network management and optimization

Effective operation and excellent performance in 5G and future networks require efficient network management and optimization. Many sources underline it. The Open Radio Access Network (ORAN) optimizes 5G networks by improving flexibility, scalability, and efficiency. Maximizing ORAN benefits and network performance requires efficient network administration. Ericsson stresses low-energy and load-dependent network management to improve 5G and beyond energy performance. Efficient network optimization reduces mobile network energy use and environmental effect. Telecom Global stressed that 5G network optimization is essential to meet rising data demands and support VR, HD video streaming, and IoT devices. Optimizers are essential for network abnormality detection, capacity management, and quality assurance [2].

Efficiency is crucial when it comes to network administration and optimization for a variety of reasons:

- Improved Performance: Networks have been optimized to meet the diverse requirements of modern applications and services, offering high data rates, minimal latency, and reliable connectivity.
- Resource Efficiency: Efficient resource management is key to optimizing resource allocation, minimizing waste, and boosting network capacity.
- Quality of Service: Network optimization is crucial for enhancing user experiences, ensuring customer satisfaction, and meeting service-level agreement compliance by upholding high service quality standards.

Optimized networks minimize energy consumption, promoting sustainability and lowering operational costs. To successfully navigate the challenges of the 6G era, it is crucial to have a well-optimized network management system that can efficiently handle various deployment scenarios, emerging technologies, and the increasing complexity of networks. Federated machine learning (ML) helps healthcare, finance, and transportation solve data privacy, confidentiality, and sensitivity issues. Federated learning (FL) lets users train models without sending data to a server [4]. This method protects privacy and security. FL protects privacy by exchanging parameters rather than raw data during training. Data privacy is addressed, and collaborative model training across numerous devices improves model performance and data security. FL can predict and diagnose cardiovascular diseases, diabetes, and cancer using varied datasets and patient privacy. As FL investments rise, organizations expect ML applications across sectors to become more efficient and successful.

FL offers a practical approach to apply ML on networks faster than 5G while maintaining privacy. FL encompasses network management, data security, privacy, and sensitivity. Without the need for data exchange or trading, several users can train the same model independently. With data from the edge device, the IoT node, and the network provider, this plan trains models. Radio resources, network slicing and automation, edge intelligence and offloading, and IoT devices can all benefit from this learning approach. FL enables you to make decisions depending on the location of the data with the aid of edge devices and IoT nodes. Your ability to learn and your response speeds both improve. Collaborative learning helps you use many datasets worry-free since it safeguards privacy [5]. More precise and dependable than they were a few years ago are network optimization technologies. In this chapter, the next section explores the fundamentals of FL. Then the applications of FL in 5G network management are explored. The final section deals with the challenges and issues of FL in 5G network management and it is concluded.

2 FUNDAMENTALS OF FEDERATED LEARNING

FL is essentially a decentralized version of ML. In conventional ML methods, models are typically trained on data that is consolidated from a variety of peripheral devices, such as smartphones and laptops, and then transferred to a centralized server. The learning process occurs on this centralized data repository, where ML algorithms such as neural networks train themselves on the aggregated data and subsequently make predictions on new data. Nevertheless, the learning methods are dispersed across the edge devices in a FL system. Rather than centralizing the training data, the model parameters are transmitted to individual devices, such as smartphones, where the learning process is conducted locally on the data of each device [6]. This decentralized ML approach offers a variety of benefits, such as the capacity to utilize insights from a variety of data sources without directly accessing the raw data, reduced bandwidth requirements in comparison to transmitting raw data, and enhanced data privacy. The emergence of open-source frameworks facilitates the development and deployment of FL architectures for training predictive models on decentralized data.

2.1 Federated learning framework

FL has been successful in many academic subjects and sectors. FL is a modern configuration for training ML models and improves the Google keyboard (G-Board for Android), predictive text, and Google Assistant. FL settings enable protected on-device ML collaboration. Personal data stay on the client's device while several clients communicate. FL uses devices for collaborative ML. Google released it in 2015. FL distributes a global ML model from a central service provider to client-edge devices. The global model is trained on each client's data on the local device to generate a trained local model and maintain client data on the client edge device. A regional model's modified parameters are sent back to the central server to be aggregated across the pool of participating clients and used to update the global model. After upgrading the international model, FL training is given to another group of consumers for a second cycle. FL is iterative and collaborative. Modular FL simplifies private and secure collaborative ML. Aggregating client local parameter changes with numerous safe computations protects client data and the global model [7].

FL is distinct from three other ML training approaches. First, it differs from traditional centralized ML training, where all local client datasets are transmitted and shared with a central server. Second, it is different from distributed ML training, which necessitates parallelization. Finally, it is different from decentralized ML training where it is assumed that the collection of datasets existing at different clients is uniformly distributed. FL is the procedure of training the ML model at the edge devices and receiving the updates from the local clients [3]. These updates are then sent to a central server since FL follows decentralized architecture as we already discussed earlier. Centralized ML starts with saving local client data on a central server and performing a training of it. Integrating the perception that FL involves unique training from the client model, the collaborative learning environment in FL improves the enhancement of the different client models to be personalized by adopting this local training that updates the global model. FL respects the privacy of the user to the extent that they can decide not to pass user data, but at most share broad model updates after many secure calculations [7].

2.2 Federated learning architectures

FL is comprised of different architectures: vertical, horizontal, and transfer learning (TL). In this section, every design is described, and the best distinguishing feature of the design is identified.

a. Vertical Federated Learning: Vertical FL divides among devices features that are interrelated but distinct and trains the model on mutually orthogonal attributes. The feature-based FL, also known as vertical FL is used in the case where multiple client datasets have overlapping sample ID but they differ in their input features. This method of learning allows clients to combine feature information for a sample ID using a third party that guarantees no information about the unique sample ID is revealed during feature sharing. Thus, vertical FL develops an ML model's cost function and gradients while safeguarding the sample ID data and collaboratively distributing its unique properties across several clients. Vertical FL expects honesty but invites questions. This assumes that an adversary has penetrated one of the clients and that the other client can only obtain data from it without being affected. Vertical FL's main benefit is encryption, which promotes privacy [8].

b. Horizontal Federated Learning: Devices or servers collaborate to train the model with data samples. Horizontal FL, or sample-based FL, is used when client datasets have different sample ID spaces but a shared feature space. For instance, two local e-commerce companies with different customer databases from different locations share a feature input space by collecting comparable user data. The two datasets or samples for a comparable service offered by the client, represented by a similar feature space, are the e-commerce shop customers. The situation above is suitable for horizontal FL setups. Horizontal FL ensures equitable information exchange across cross-users in different clients to generate a larger dataset for the artificial intelligence (AI) model. The horizontal FL may be used in a multi-tasking system to let various clients to learn different jobs while sharing knowledge between samples and protecting dataset privacy. Horizontal FL prevents data leakage during customer sharing, protecting data. Researchers said horizontal FL's main benefit is security-focused learning across customers.

c. Federated Transfer Learning: This framework optimizes pre-trained models using decentralized data for specific tasks, reducing local data. Two client datasets with a limited sample ID space but separate feature spaces employ federated transfer learning (FTL). It functions like vertical FL but with a smaller sample ID overlap. Vertical FL only works for crossing dataset sample space. TL in FTL helps you learn across the whole dataset even if there is only a tiny overlap in a similar sample space. Head retraining usually uses two datasets: a source dataset that trains an initial model with a feature extractor and a prediction head, and a target dataset that trains a new prediction head using the feature extractor from the source dataset. Using a pre-trained feature extractor on the target dataset can increase ML model accuracy. Target datasets often have fewer data samples than rich source datasets. Thus, FTL allows information sharing across clients with limited overlapping sample regions, such as users, to improve models [7].

2.3 Advantages of FL over traditional ML

In traditional ML, data is usually gathered and stored in a central location. The model is then trained using this central dataset, which involves transferring sensitive data between a cloud location and devices. On the other hand, as previously mentioned, the FL platform operates by transferring models to the data rather than moving data to the models. One notable distinction between federated and classical ML is that FL often entails training models in a distributed manner, where each server or device plays a role in the training process. In contrast, traditional ML techniques involve training models on a single machine owned and operated by a specific resource.

a. Improved data privacy and security

   FL's collaborative nature allows it to treat each IoT device at the edge as a unique client, enabling models to be trained without transmitting raw data. During the FL process, each IoT device only collects the necessary information to fulfill its task. With FL, raw data is stored on the device and only model updates are sent to the central server [7]. This can prevent leakage of personal raw data, thereby ensuring safe operations.

b. Reduced costs and cost per unit of output

   As a consequence, it has been determined that one more important aspect in data science is the availability of greater amounts of data. For instance, training the model with big data can be iterative and time-consuming as opposed to having all the data consolidated in a single server or cloud. While in the FL the datasets do not necessarily have to be stored together and the models can be trained at different data points at an instance. This greatly speeds up the deployment of AI and IoT models and enhances the scalability of operations. In addition, when FL trains the model at the edge and only transfers the trained weights, it significantly decreases the amount of data that needs to be sent over the network and processed in the cloud. By doing so, it helps to reduce expenses related to data transfer, computing, and storage [9].

c. Enhanced flexibility

Typically, FL models can adapt to new scenarios without requiring retraining. These models can leverage past experiences to make predictions. In addition, the skill and experience that could be attained through this learning method can be applied to other areas of study. For instance, where the model is conversant with environments or settings, it can employ that knowledge in another environment or setting to make the operations more efficient, less costly, and faster.

d. Enhanced precision and variety of data

Whenever an organization seeks to train its model with data, there is a possibility that the data used may not encompass a bigger dataset that the model will be dealing with. But, it also means that when trained on data from different source, it encounters a wide range of additional information, the way of "seeing" the new data reduces the variation and its capacity to adapt improves.

3 FEDERATED LEARNING APPLICATION IN 5G NETWORK MANAGEMENT

In the next section, we will expand on the application of FL in 5G network management.

FL holds several usability promises in the field of 5G network management as discussed below. Here are some important areas where FL can be utilized:

a. Network Optimization and Resource Allocation: FL could be utilized to cooperatively train models that can enhance the network's resources; these include bandwidth, load, and cell. It can also consolidate data from dispersed base stations and user devices to create complex algorithms that can effectively organize resource utilization and network traffic [10].
b. Mobility Management and Handover Optimization: Symbolically, 5G networks should offset the moderate mobility of users from one cell to another and access point. Handover and mobility can be

forced in FL to train predictive models which may help in predicting mobility patterns or choices for a given user. This is done through data gathered from various network elements and user devices that are involved in the call establishment and connection processes.

c. Network Slicing and Virtualization: Another key element implemented on 5G networks is the use of network slicing and virtualization to provide unique services that fulfill the needs and desires of different applications and users. FL can be applied in the training of models for the dynamic management and control of the network slices where conclusions are drawn from distributed data from the network elements and end-use devices.

d. Network Security and Anomaly Detection: FL can enable cooperation regarding model training to detect security threats, anomalies, and cyberattacks in the 5G networks. FL can strengthen security measures by collecting data from dispersed network constituents and user gadgets and without compromising the privacy of the data.

e. Self-Organizing Networks (SON): It depends on the fact that due to versatile conditions and diversified user requirements, 5G networks must have the self-organization and self-optimization feature. FL may be used to train models that are for system optimization and network such as self-configuration, self-optimization, and self-healing. This is achieved through data from the dispersed network elements and the user device information.

f. Management of QoS and Quality of Experience (QoE): FL may be applied to train prediction models that can improve the desired QoS and QoE parameters based on information received from user devices and network components. This can improve the reliability and QoS for various uses and concepts in 5G systems.

g. Energy Efficiency and Green Networks: FL may be utilized to create energy-efficient algorithms and models for 5G networks by utilizing data from dispersed network elements and user devices. This can aid in optimizing energy usage and reducing the carbon dioxide emissions associated with 5G networks.

h. Core Edge Caching and Computing: Processing of data and caching of contents at base stations reduces backhaul usage, thus reducing burden on this aspect. Popular material should be routed closer to SBSs, access points, and user devices for faster local access. This paradigm may enable high-latency, high-bandwidth apps. Research is ongoing on which contents to store in each cache to make this architecture work. The literature uses static or dynamic statistics to determine content popularity. Dynamic methods indicate immediate popularity using content statistical features, while static models disregard time-varying popularity. Model-driven content popularity detection cannot account for all popularity causes [11]. FL, which replaces user input with locally trained models, is ideal for wireless network proactive caching content popularity prediction. FL teaches AR users fundamental augmentations without collecting personal data. Popular content is pre-fetched and cached locally to decrease latency. FL and pre-cached roadside devices allow self-driving cars to access traffic data from other vehicles [12].

i. Spectrum Management: Millimeter-wave (mm-wave) propagation physics allows for spectrum access regulatory reform. Future 5G networks will need low band and high band frequencies in the microwave and mm-wave bands with different licenses to support important 5G verticals. For optimal 5G network performance, collaborative and autonomous spectrum sharing solutions customized to the environment and applications are essential. However, dynamic and dispersed spectrum access is complicated. Privacy issues make accessing and sharing all radios' high-resolution spectrum use statistics difficult. All radios must communicate sensory data including spectrum occupancy, device nonlinearity, and interference detection like database administrators. However, radio operators may not divulge their frequencies due to privacy concerns. Spectrum access databases and centralized spectrum utilization data collection systems may not work. Processing big datasets needs sophisticated processing and extensive optimization, which can be computationally difficult. Thus, spectrum autonomy requires crowdsourced, decentralized intelligent radio networks that share spectrum [13]. To overcome these concerns, federated ML may communicate radios local spectrum usage models. The aggregator delivers radios a global model updated with local spectrum utilization model parameters for spectrum access decision-making. Two wireless systems can share a technique. This method may help integrate DSRC and c-V2X in the same ITS band.

j. 5G Core Network: The new 3GPP network data analytic function (NWDAF) improves data exposure for core network ML-enabled operations. It allows smart network management methods. Automating network administration and setup reduces operational expenses by removing human–machine interface. NWDAF may use any basic network data and connect to any NF. Additionally, any NF can request network analytics. Back to the federated setup, we found that the datasets had the same feature space while being based on various stages of the process (sample spaces). Data dissemination like this is horizontal fragmentation. Vertical fragmentation occurs when distributed datasets share sample space but differ in feature space. Vertical federated ML for fragmented datasets utilizing local learners was prompted by data fragmentation. Vertical federated ML uses feature space for non-IID, imbalanced, and enormously scattered datasets. Vertically fragmented datasets will be compared using two datasets with all network subscribers and the same sample space. Features may vary. One dataset may contain registration and authentication data, while the other may provide user network slice selection data. Vertical FL works well in a core network where each entity handles user-relevant dataset attributes. The AMF and SMF manage mobility, session formation, IP address allocation, and traffic routing. NWDAF may collect user data worldwide. Each core network item holds one feature for all users in vertically divided databases. Vertical FL lets core network entities submit encrypted models based on locally derived data characteristics instead of raw data to the NWDAF unit. This can significantly minimize network topology network function virtualization (NFV) cybersecurity threats [14].

k. IoT and mMTC management: With the advent of FL, a world of possibilities emerges for streamlining the handling and enhancement of IoT and mMTC in 5G networks. Given the rapid growth of IoT devices and the rising need for mMTC services, it is imperative to implement robust management strategies to guarantee dependable, protected, and streamlined operations. When it comes to managing IoT devices, FL can be extremely valuable in maximizing resource allocation, offloading tasks, and improving battery life. Due to the shared training it allows a cooperative learning where multiple models across different geographical locations/organizations make use of the data from the IoT devices to improve conditions depending on scenarios, communication protocols, data compression techniques, and security measures. Thus, these models can be further updated and optimized when new data is received, ensuring that IoT devices perform optimally as well as operate securely, without compromising users' data privacy. This work also points to that FL practices are promising in interpreting IoT data and deploying methods of predictive maintenance. Especially when it comes to approaches like anomaly detection, condition monitoring, or even predictive maintenance, the use of the distributed training mechanism can become a valuable solution that will improve the overall accuracy and reliability of the models. It also helps in ensuring inside data protection by making sure that only authorized personnel get access to the data. In this way, with the aforesaid cooperation at their core, the main concerns can be identified in the early stages, thereby promoting the minimization of the system's downtime, as well as the extension of IoT equipment's life cycle. This approach enhances reliability of the IoT systems and the overall experience within the system.

In the case of mMTC, FL has the features with the possibility to improve traffic management and scheduling 5G networks. Managing a significant amount of low-power tier devices that have an on-off connection and stochastic traffic characteristics is not easy to solve since it is significantly different from managing nodes that have always been connected, constantly, and homogeneously. It is noted that FL can be used to train models for resource allocation, traffic scheduling, and interference management with data collected from multiple base stations and devices. They can adapt to new environments and even make micro decisions such as increasing the capacity of the transmission power, tilt angle, and frequency band. This makes it certain that key mMTC application can have stable and sufficiently quick communications with without much latency. Moreover, FL can also help in improving the links and coverage for mMTC devices in the network. The core of FL is the capability to learn models based on data arising from distributed base stations, user devices, and environmental sensors, thus capable of adapting to new conditions and adjusting finer details such as transmission power, antenna configurations, and frequency allocation. This way, it is possible to enhance the QoS and QoE for mMTC applications while taking into consideration the load balance between the end device, MBS, and PBS. This will help people to be understood rightly in various situations and for there and then to get right communication strategies.

To effectively manage mMTC, there is a critical need to maintain adequate security measures, and in this regard, FL offers potential in modifying the security and authentication standards significantly. According to FL, security threats can be identified during training, as well as authentication and access control can be established with the use of collaborative training. This approach allows the integration of data from mMTC devices, base stations, and security systems while keeping the data secured without leaking. These models provide adequate security and protection mechanisms that help in protecting mMTC applications and its communication channel against various threats. However, as with most ML federations, there are some challenges that need to be addressed in order to take full advantage of FL in regard to IoT and mMTC. The main important issues are how the models exchange information with each other, how the participants are motivated to provide data, and how to securely aggregate the new updates to the models. Further research will be devoted to these issues, thus aiming at the ultimate inclusion of FL into the IoT and mMTC handling in the 5G environment. Distributed and decentralized FL works in coordination with the distributed and layered structure of 5G networks to achieve collaborative learning as well as optimization of models while preserving the data privacy and ownership. However, in order to implement FL into the management of the 5G network successfully, there are several challenges that need to be addressed such as challenges touched on efficiency of communication between participants, incentive schemes, and safe communication aggregation.

4 CHALLENGES AND FUTURE DIRECTION

Most investigations related to FL are in the preliminary stage. It is important to note that there are advantages that could be derived to the adoption of FL in wireless networks; there are also enormous challenges that need to be taken care of. Finally, there will be a brief discussion about the future research direction and some of the challenges associated with the current system.

4.1 Security and privacy challenges and considerations

Another reason for federated ML is the local datasets' privacy; therefore, the data would remain private in the local domain. It is notable that a secure aggregation algorithm has been used to allow the aggregation of the encrypted local models without the necessitation of decrypting them within the aggregator in a way that models end up point to their data.

However, if one has to look at the whole picture of model, the participation of a particular learner from the local context can be disclosed. The idea proposed here is to offer the privacy at the local learner-level, not for a single data sample as is the case normally. However, these algorithms might negatively affect the general performance of the model or require extra computational hardware resources, or a particular number of local learners to be engaged in building up the model [12]. Hence, there is a need to develop efficient federated algorithm that performs well to achieve model federations and protect privacy of data without escalating complexity. Moreover, there exist some certain types of the neural network models that contain the definite features of the training examples. This goes without saying in cases where models are trained with the help of sensitive user data, which is the case with FL, for example. Data memorization and its optimization play a significant role in federal learning to get the benefit of user data without leaking any data when the attackers launch an attack [15]. As in the other ML methods, in FL, local models go through new data collected and are retrained to update the trained model. Therefore, an adversary has the possibility to control the local training datasets without being observable by the other parts of the system with the aim to affect the result of the model. This is accomplished by deliberately incorporating certain samples to act as poison that will compromise the FL process. The latter has the potential to become a threat to the model since it has the possibility to send gradient updates with the purpose of implementing a model-poisoning attack [12]. An adversarial lens has also been applied toward analyzing FL and identifying the ability of the learning process as a target for model-poisoning adversaries. It is imperative to build strong defenses as the primary style of for FL is under threat by adversarial attacks. Moreover, an aware aggregator or a close student has the potential to commit a membership inference attack on other close students. Consequently, the crime of an inference attack is to identify whether a given point belongs to a training dataset. Adjusting the parameters of the model is critical since it enables the corresponding improvement of the membership attacks' precision. Inference attacks can be in various types, and all these may lead to a threat to the privacy of the local learners. The following are the major types of attack: attributive attack, which comprises parameter inference, input inference, and attribute inference [12].

FL may allow performing collaborative learning and optimization of 5G network management and its parameters based on examples while preserving the data. Nevertheless, as with most paradigm shifts, there are several hurdles and future prospects that must be addressed in order to effectively deploy and adopt FL in this industry [16]. Many key challenges and prospective development pathways are listed below:

1. Communication Efficiency and Overhead: In the FL method, the central server and devices or other network nodes provide and receive model updates. There are likely to be some overheads associated with this connection in particular if perhaps bandwidth or the device in question possesses a small amount of resources. Model compression, on the other hand, is critical for FL systems because it helps in reducing the communication overhead and increasing efficiency via quantization and elimination of redundancies.
2. Heterogeneous Data and Non-IID Distributions: 5G networks collected heterogeneous information that is not limited but independently and identically distributed (non-IID) across devices or network partitions. This can make convergence more difficult and negatively affect the outcome of FL models. More attention needs to be paid to the development of efficient FL algorithms eliciting acceptable performance on the non-IID data types.
3. Incentive Mechanisms and Fairness: FL latter demands equal share of device or network involvement Usually, FL lacks equilibrium of device or network contribution. The incentive structures need to promote participation and at the same time dissuade those who will only opt to "take the free ride" and those who have ill-intentions. Fairness issues are necessary to avoid trained models from showing biases or discriminating against user groups or places.
4. Secure Aggregation and Privacy Preservation: FL protects data privacy by not sharing raw data, but model updates or inference attacks can compromise privacy. Secure multi-party computing, differential privacy, and homomorphic encryption are needed for 5G network management aggregation and privacy.
5. Scalability and Distributed Learning: As 5G networks grow, FL systems must manage many devices and network portions. To scale and optimize learning in massive 5G networks, distributed learning architectures, parallelization methods, and hierarchical FL must be explored.

6. Federated Transfer Learning and Model Personalization: FL may enable collaborative learning and TL for devices, network components, and geographic areas. Effective TL and customization procedures must be established through FL framework research to optimize global and local model performance.
7. Network Slicing and Virtualization: 5G networks provide tailored services and resource allocation based on application and user demands. FL methods must be integrated with network slicing and virtualization frameworks to optimize resource allocation, service optimization, and security across network slices.

Standardization and interoperability need to be applied to 5G network management to utilize FL. There are more fundamental requirements for convergence on protocols and methods, however, in order to assimilate different vendors and operators. There is no doubt that FL frameworks, protocols, and interfaces should be the direction of the development of standard work-streams and industry collaborations in 5G networks.

4.2 Challenges and considerations related to the algorithm

As aforementioned, the convergence of an algorithm is an essential aspect, especially for decentralized algorithms such as FL, when communication and computation resources are scarce. In the following the analysis for determining the convergence rate of gradient descent-based FL for convex loss function has been made. The following holds useful to perform analytical assessments on the circumstances that result in the convergence of the algorithm for non-convex loss functions. This is particularly relevant in some models such as deep neural networks, and the aim is to learn non-convex mapping. However, the following are some of the factors which one should consider in order to arrive at the correct procedure of updating the global information security. These are the desirable number of local learners to include, the manner of categorizing them, and the frequency of download locally and globally. These decisions concern such factors as when is enough in terms of model capability on one side and the resources used on the other, and these are different for various contexts. In addition, some ideologies such as federated deep neural networks may still contain information updates that cannot be stored in low-powered device such as IoT nodes. Thus, methods that help to simplify and concisely represent the model parameters are extremely effective in such a process a computation and usage of resources.

4.3 Challenges and considerations in wireless settings

The limited bandwidth available in the wireless systems requires the information signals to be quantized when it is transmitted across the channel. Since the locally updating learner and the aggregator require the exchange of model parameters across the wireless channel, this culminates in FL with parameter quantization. Another aspect that can be explored in this context is the impact of quantization error on the stability of models during training. Besides the transmission capacity, the other parameters that keep on aggravating the channel bottleneck are noise and interference. This is a critical function because it assesses the capacity to handle these channel effects [6]. Another parameter that must be considered is the time required to convergence in case of using this method. The FL convergence time comprises the computational time that each local learner and aggregator take and the time it takes for the two to exchange data over the wireless channel. So, when defining the degree of frequency of local updates and global aggregation, it is crucial to consider the quality of the wireless channel. Furthermore, when in the process of learning on a deep model, there are usually techniques such as model compression which can be used to reduce the complexities of the model and practices such as sparse training which can also be used to reduce the number of model parameters [17]. These methods are advantageous for devices that do not possess a large computational ability for obtaining such models. This way, there is always a way to make the process less complex without necessarily losing too much accuracy of the model. As such, in the process of optimizing federated deep learning models applied to wireless applications, the costs of communication and the quality of the wireless channel have to be considered. Moreover, since the wireless channel "availability" depends on the wireless channel quality, it may be possible to compress the model in a dynamic way, according to the wireless channel quality. However, since competence, availability of devices, and desire of learners to be involved in the learning process are other significant determinants of selection for training, the quality of the wireless channel between the global aggregator and a specific local learner will also influence the learning process. This element should also be discussed in conjunction with other criteria in the decision-making process. At times, a certain device may want to input its data into the model, but the wireless channel available for the device might not have enough bandwidth to allow the model parameters to be transmitted with adequate quality. This might lead to some inaccuracies of the global model; this could lead to some shortcomings on the part of the global model [18,19].

Several open issues and challenges in network management can be addressed through FL, including collaborative learning, optimization, and privacy protection during the accommodation of 5G networks and new applications. As for approaching the future actualization of FL in the 5G network management, these issues should be solved and corresponding further solutions should be considered.

5 CONCLUSION

FL is closely associated with several crucial 5G administration aspects and proves to be a very promising technique. Particularly, it frees up the flow of learning and cooperation and guarantees the optimization of the joint processes, as well as ensuring the protection of data and their complete sovereignty. This approach toward distributed ML makes it possible for so many devices to contribute to training of the same model simultaneously but the individual data does not have to be exposed to the rest of the devices. One of the massive advantages is in the area of privacy with regards to data, expandability, and cost-effectiveness. Thus, this chapter aims to explain several components that define FL: the structure and a few primary categories of FL, including vertical, horizontal, and FTL. Moreover, it elaborates how FL is better than conventional ML or other forms of learning methods since it offers better data protection and safety, and higher flexibility, and it is cost-effective in terms of computing costs. FL in 5G network management uses is very extensive, and it has numerous impacts on one or several aspects of a particular society. They are used in such areas as routing, networking, data security and control of IoT devices at various stages. It also implies that the employment of FL may also improve the reliability and effectiveness of the IoT systems to achieve superior QoS and QoE for applications based on mMTC. In addition, FL can be a game changer for improving security and user authentication for mMTC devices. However, the application and employment of FL in the management of 5G networks present many considerations such as security and privacy, communication, and the all-embracing heterogeneity of the data. In order to address these problems, it is necessary to refine algorithms at the academic and developer level further, as well as implement safer aggregation strategies and concrete protection measures. Thus, management of the 5G networks using FL has benefits in terms of optimization and security which has been discussed in this chapter. As for further research in this field it is inevitable to note its importance for the effective implementation of application of the discussed concept.

REFERENCES

1. W. Mwashita and M. O. Odhiambo, "6G and D2D as Enablers of the Convergence of Artificial Intelligence (AI) and Internet of Things (IoT)," *Transactions on Computational Science and Computational Intelligence*, 2024, pp. 21–41, doi: 10.1007/978-3-031-55718-7_2.
2. "Overview of 5G Technology and Its Applications," *element14*. https://vn.element14.com/5g-what-is-5g-capabilities-and-its-applications?ICID=I-CT-TECH-RES-CLA-FEB_22-393
3. G. Cisek and T. P. Zielinski, "Prototyping Software Transceiver for the 5G New Radio Physical Uplink Shared Channel," Sep. 2019, doi: 10.1109/SPS.2019.8882079.
4. "Understanding Important 5G Concepts: What Are eMBB, URLLC and mMTC?," *Featured News Story|Verizon*, Jun. 05, 2023. https://www.verizon.com/about/news/5g-understanding-embb-urllc-mmtc
5. M. Liu, R. F. Yu, Y. Teng, V. C. M. Leung, and M. Song, "Mining Task Offloading in Wireless Blockchain Networks," in *Springer eBooks*, 2020, pp. 850–852, doi: 10.1007/978-3-319-78262-1_130.
6. M. Gupta, P. Sharma, and R. Kalra, "Federated Learning and Artificial Intelligence in E-Healthcare," in *Advances in Healthcare Information Systems and Administration Book Series*, 2023, pp. 104–118, doi: 10.4018/979-8-3693-1082-3.ch006.
7. S. Banabilah, M. Aloqaily, E. Alsayed, N. Malik, and Y. Jararweh, "Federated Learning Review: Fundamentals, Enabling Technologies, and Future Applications," *Information Processing & Management*, vol. 59, no. 6, p. 103061, Nov. 2022, doi: 10.1016/j.ipm.2022.103061
8. S. Choudhury, "Architecture and Design Considerations for Next-Generation Mobile Networks with Integrated Edge Computing," *rucore.libraries.rutgers.edu*, 2024, doi: 10.7282/t3-8qqm-tf43.
9. O. A. Wahab, A. Mourad, H. Otrok, and T. Taleb, "Federated Machine Learning: Survey, Multi-Level Classification, Desirable Criteria and Future Directions in Communication and Networking Systems," *IEEE Communications Surveys and Tutorials/IEEE Communications Surveys and Tutorials*, vol. 23, no. 2, pp. 1342–1397, Jan. 2021, doi: 10.1109/comst.2021.3058573.
10. H. Li et al., "Review on Security of Federated Learning and Its Application in Healthcare," *Future Generation Computer Systems*, vol. 144, pp. 271–290, Jul. 2023, doi: 10.1016/j.future.2023.02.021.
11. P. M. Tshakwanda, "Enabling Intelligent Network Management through Multi-Agent Systems: An Implementation of Autonomous Network System," *UNM Digital Repository*. https://digitalrepository.unm.edu/ece_etds/630
12. S. Niknam, H. S. Dhillon, and J. H. Reed, "Federated Learning for Wireless Communications: Motivation, Opportunities, and Challenges," *IEEE Communications Magazine*, vol. 58, no. 6, pp. 46–51, Jun. 2020, doi: 10.1109/mcom.001.1900461.

13. B. T. Geetha, B. Vidhya, K. Pushpalatha, and M. Priyadharshini, "A Comparitive Study of 5G Technology with Other Generations of Mobile Technologies," *2017 IEEE International Conference on Power, Control, Signals and Instrumentation Engineering (ICPCSI)*, Sep. 2017, doi: 10.1109/icpcsi.2017.8392020.
14. Q. Yang, Y. Liu, Y. Cheng, Y. Kang, T. Chen, and H. Yu, *Federated Learning.* 2020, doi: 10.1007/978-3-031-01585-4.
15. Yadav, S. P., Bhati, B. S., Mahato, D. P., & Kumar, S. (Eds.). 2022. *Federated learning for IOT applications*. Springer International Publishing.
16. G. Singh, "5G principles," *Telecom Trainer*, Dec. 14, 2023. https://www.telecomtrainer.com/5g-principles/
17. I. Gupta and P. Pathak, "Cybersecurity in Digital Epoch: Emerging Threats and Modern Defense Techniques," *AIP Conference Proceedings*, Jan. 2022, doi: 10.1063/5.0109715.
18. T. M. Mengistu, T. Kim, and J.-W. Lin, "A Survey on Heterogeneity Taxonomy, Security and Privacy Preservation in the Integration of IoT, Wireless Sensor Networks and Federated Learning," *Sensors*, vol. 24, no. 3, p. 968, Feb. 2024, doi: 10.3390/s24030968.
19. T. A. Kumar, A. Gokulalakshmi, P. Kanimozhi, and G. Glorindal, "Application Domains of Federated Learning in Healthcare 5.0," in Ahdi Hassan, Vivek Kumar Prasad, Pronaya Bhattacharya, Pushan Dutta, Robertas Damaševičius (Eds.), *Advances in Healthcare Information Systems and Administration Book Series*, IGI Global, Pennsylvania USA, 2023, p. 391, doi: 10.4018/979-8-3693-1082-3.ch016.

CHAPTER 6

Non-orthogonal multiple access wireless systems using deep learning

Rudraksh Gohil and S. Deepa

1 INTRODUCTION

1.1 Background of Non-Orthogonal Multiple Access

Non-orthogonal multiple access (NOMA) is an advanced, multiple-access technique that has developed recently to increase the efficiency as well as the capacity of wireless communication systems. Although in conventional orthogonal multiple access (OMA) techniques various resources have been assigned differently to serve different users, via NOMA, a system can simultaneously allocate identical time, frequency, and code domain resources to serve multiple users. This is achieved through the use of overlaying different power-level signals and then decoding them through interference cancellation at the receiving end.

The primary interest behind the technique is to improve spectral efficiency and connectivity for users. With the concept of allowing multiple users on a given resource block, NOMA can significantly increase the number of users that can be served at any instant, thereby increasing the overall capacity of a system. It is instrumental in heterogeneous

DOI: 10.1201/9781003514336-6

user requirements and channel conditions, which characterize 5G and beyond networks.

Users' basic service principle in NOMA systems indicates their potential to tolerate high-level interference in the event of a good channel condition relative to each other. As such, power allocation in NOMA systems becomes critical since those users with successful strong channel conditions are served low power, and those who are weak in that end are served with high power. This power domain multiplexing ensures that the receivers can effectively separate the signals using successive interference cancelation, where the signal with the highest power is first decoded, and then its contribution is subtracted from the combined signal before the next one is decoded.

1.2 Importance of NOMA in 5G and beyond

5G of mobile networks aims to meet the ever-growing demand for higher data rates, lower latency, and massive connectivity. To achieve these goals, 5G leverages various advanced technologies, among which NOMA plays a crucial role. NOMA contributes to 5G's objectives by:

a. **Increased capacity:** NOMA holds considerable potential for furthering 5G's goals since it increases the capacity available. Multiple users may share the same resource block, thus increasing the number of users served at any given duration and enhancing network capacity, which is very important in dense urban environments where user demand is high.

b. **Improved spectral efficiency:** More than one signal can be overlapped on the same spectrum through NOMA; hence, it significantly enhances the spectral efficiency of the resources. The efficient use of the already congested resources of the spectrum is the paramount requirement for the next-generation communication systems.

c. **Higher user fairness:** NOMA can dynamically assign power levels to different users depending on their channel conditions. This will help in maintaining the fairness of the resources among the users in a heterogeneous network, where the data rate requirement may be different for different users and their channel conditions are distinct.

d. **Flexibility:** It can be complimentary with other technologies of 5G like massive multiple-input multiple-output (MIMO), communication in millimeter wave, and beamforming to give further performance enhancement of the network. Such capacity creates rigid and flexible communication systems that meet different user requirements and environmental conditions.

e. Reduced latency: NOMA reduces the overall latency in the network, considering it allows simultaneous transmission to multiple users. The former will be critical in applications with real-time communications, such as autonomous driving, remote surgery, and virtual reality.

Going beyond 5G is expected to keep further alive the quest for more efficient and scalable communication systems. With its capability to serve many users with different demands and requirements, NOMA has once again shown up as one strong candidate for future network technologies, possibly bringing the benefits up to 6G and beyond. Beyond 5G networks, the integration of NOMA with emerging technologies such as terahertz communication and the ultradense network can achieve further enhancement of the system's capacity and efficiency.

1.3 Basic introduction to deep learning and its importance in NOMA

Deep learning (DL) is an exciting subfield of machine learning (ML) that has recently taken the spotlight due to its focus on modeling complex patterns and making intelligent decisions. In particular, DL algorithms like neural networks have demonstrated state-of-the-art performance in various fields, such as computer vision, natural language processing, and, more recently, wireless communications.

When embedded within NOMA, classical DL will further enrich the ultimate performance of the wireless network. DL can apply to the following aspects of NOMA:

a. Channel estimation: This is especially important in implementing NOMA. DL models may exploit historical data of channels to forecast the future states of the channels and enhance the overall performance of NOMA systems.

b. User clustering: Users could be clustered in a NOMA system considering the heterogeneous nature of the channel conditions between the users. DL, by nature, captures real-time user clustering, including mobility patterns, channel conditions, and traffic demands.

c. Interference management: Interference management in NOMA systems is a chief concern. DL algorithms can be applied to predict interference patterns, optimize power allocation, and beamforming strategies to mitigate the same, hence paving the path for seamless communication by any user.

d. Resource allocation: Effective resource allocation is necessary for NOMA systems to maximize performance. DL algorithms are utilized to maximize system performance and user satisfaction by optimizing resource allocation among power bandwidth and time slots. In a dynamic setting, the algorithms adjust quickly and nimbly to user needs and shifting network conditions. DL underpins adaptive modulation and coding which modifies the schemes in response to user demands and channel conditions. This guarantees each user the best possible data rate and dependability.

2 NOMA IN 5G AND BEYOND NETWORKS

2.1 Role of NOMA in 5G

The importance of NOMA in enhancing the performance and efficiency of 5G networks cannot be overstated. NOMA allows multiple users to share the same time frequency and code resources, thus overlaying user signals with different power levels for simultaneous transmission. This means that one can stack several layers of user signals on top of each other at varying intensities so as to transmit them concurrently. In order to achieve better spectrum efficiency in addition to connectivity for more users in 5G with high data rates, NOMA should be used. It supports various applications such as Internet of Things (IoT) and enhanced mobile broadband (eMBB). Ultra-reliable low-latency communication (URLLC), among others, is also supported by this technology. By employing successive interference cancellation (SIC) at the receiver end, NOMA can separate and then decode signals meant for different users from each other. In this context, major benefits associated with NOMAs vis-à-vis 5G include fairness improvement among users, higher throughput rates at reduced latencies as well as increased spectral efficiency, especially when dealing with massive MTC (machine type communication) devices under diverse channel conditions where conventional OMA may not work efficiently due to the large number of users involved. It is important for 5G networks to achieve very aggressive performance targets, and this is where NOMA comes in, as it does not require extra bandwidth to serve many clients simultaneously.

2.1.1 Enhanced spectral efficiency

By providing multiple users with access to the same time and frequency resources, NOMA improves spectral efficiency. This principle is implemented through power domain multiplexing where signals are superimposed at different power levels. The stronger channels get lower power signaled while weaker ones get higher powered signals. SIC at the receiver allows each user's signal to be correctly decoded even if their frequencies overlap. The ability of NOMA to improve spectral efficiency is crucial in

5G systems that operate within a limited spectrum resource which is very valuable. The use of NOMA leads to efficient exploitation of this spectrum hence supporting ever-increasing demands for high-speed data services such as HD video streaming, virtual reality, and cloud computing among others [1].

2.1.2 Increased capacity and connectivity

One of the major objectives of 5G is to support a huge number of connected devices commonly referred to as the IoT. In this, NOMA has an integral role because it increases the number of users that can be served concurrently by significant margins. This is particularly advantageous in dense urban areas and environments with high population density like stadiums, concerts, and smart cities. The overall system capacity of NOMA is increased by enabling multiple subscribers to share the same resource blocks. This is important for accommodating the growth in the number of IoT devices predicted to reach billions shortly. By efficiently distributing resources among many devices, NOMA assures reliable and consistent connectivity even under heavy network congestion.

2.1.3 Reduced latency

It reduces the latency of 5G by NOMA because it allows for concurrent transmission to multiple users within the same time-frequency resources. When serving users sequentially with traditional OMA schemes like time-division multiple access (TDMA), higher latency is achieved. On the other hand, NOMA's power domain multiplexing supports simultaneous transmissions that allocate different power levels to individuals depending on their channel conditions. This means that each person's data is sent at once, thereby minimizing waiting time, which greatly decreases overall system latency. Additionally, NOMA can adjust resources in real time, enabling networks to become more flexible while also improving its transmission process, hence lowering latency. Such capability becomes necessary, especially in 5G systems where there are applications sensitive to delays such as autonomous driving and augmented reality or even live streaming videos.

2.2 Benefits and challenges of NOMA implementation

2.2.1 Benefits

a. Enhanced user fairness: NOMA assigns power levels based on users' channel conditions to ensure that resources are distributed fairly.

This flexibility ensures quality of service (QoS) is provided to all users, even those with poorer channel conditions. NOMA guarantees a more equitable allocation of resources so that every user gets a suitable level of service, regardless of where they are or the quality of their channel.

b. Flexibility in integration: Massive MIMO, millimeter-wave communications, beamforming, and other cutting-edge 5G technologies can all be used in combination with NOMA. Combining NOMA with multiple technologies can result in even better network performance through this synergy, for example, by increasing system capacity overall as well as improving spatial diversity through massive MIMO with NOMA.

c. Scalability: For scalability purposes, NOMA can handle high numbers of users which makes it applicable for future networks. It meets the growing demand for connectivity beyond simply 5G networks. As the number of connected devices grows, the scalability aspect of NOMA will be crucial in maintaining effective and reliable communication in increasingly crowded network environments. In achieving flawless operation and connectivity this is vital as technology advances over time.

2.2.2 Challenges

a. Interference management: One of the major challenges of NOMA is managing interferences between co-channel users. Reliable communication depends on accurate channel state information (CSI) and efficient interference cancellation techniques for reducing interference. Interference management becomes more complicated as the number of users sharing the same resources increases.

b. Receiver complexity: When SIC is used, the receiver is made more complex. The correct decoding of the useful information necessitates an error-free estimation of interfering signal power and channel attributes by a receiver. Consequently, dependable technology that can satisfy SIC's computational requirements in addition to advanced signal processing techniques is essential.

c. Power allocation: Optimizing power distribution among multiple users with different channel conditions poses a challenge. Inappropriate power allocation could lead to uneven resource sharing among users and poor performance levels. Efficient dynamic power allocation algorithms need to be developed to get the best out of NOMA.

d. Security issues: Sharing resources by many users can have security problems due to the need to keep NOMA's advantages, it is necessary to use very tough encryption and security measures for maintaining security in communication of this kind. To keep user data secure and free from eavesdropping, defenses must be set up to prevent illegal access. Therefore, there should be strict user authentication and control mechanisms put in place.

2.3 Future prospects of NOMA in beyond 5G networks

2.3.1 Integration with emerging technologies

It is possible to make use of NOMA with leading-edge processes of communication such as terahertz communication, ultra-dense networks, and intelligent reflecting surfaces in the forthcoming 5G network. The bandwidth resources that are offered by terahertz communication can be large and when combined with NOMA, it can significantly improve data rates and capacity. What ultra-dense networks offer its beneficiaries are small cells packed closely together just like a pack of sardines, as it allows NOMA to efficiently distribute its resources among several users within the same area. Moreover, intelligent reflecting surfaces adjust the dynamical propagation environment to enhance signal quality and reduce interference in NOMA systems. This could help further develop more advanced capabilities for NOMA using these new technologies hence solving future wireless network challenges. For instance, holographic communication or advanced virtual reality needs high data rates; therefore, NOMA combined with terahertz communication will allow an extremely fast transmission of information which is required for supporting such applications.

2.3.2 Artificial intelligence-driven network optimization

Artificial intelligence (AI) and ML application in network optimization for the future of 5G+ networks is expected. Power allocation, user clustering, and interference management can be dynamically adjusted by DL algorithms to optimize NOMA performance. This means that AI-driven NOMA systems would be capable of understanding real-time network conditions and user behavior leading to effective communication which learns over time. Moreover, it uses some mechanisms of AI and ML to identify potential threats and mitigate them promptly, thereby making the security and robustness of the NOMA system stronger. Ongoing network traffic analysis together with anomaly detection allows proactive security response as well as ensuring data integrity in AI-controlled NOMA systems.

2.3.3 Enhanced user experience

Future networks will prioritize not only connection and ability but also the general experience of the user. This is in line with the objective that

NOMA can offer a large number of subscribers without considering their channel conditions, high quality, and uninterrupted service. By letting AI collaborate with NOMA on top of using advanced technologies, the user experience could be improved even further by ensuring smooth communication all the time. For instance, when we have remote healthcare applications, smart city infrastructure as well as immersive gaming among others, enhanced user experience is particularly important for these applications that need reliable and good quality services. However, its ability to deliver uniform performance in different network environments makes it a useful technology for such applications.

2.3.4 Research and standardization efforts

To deploy NOMA beyond 5G networks, ongoing research and standardization efforts should be complementary. Consequently, this will lead to novel NOMA methods, optimization techniques, and security approaches by the collaborative work of academia, industry, and standardization bodies. This will be done through standardization to ensure the broad adoption and interworking of NOMA technologies across devices and network infrastructures. Setting common standards and protocols will not only bring faster deployment of NOMA, but also the integration will be flawless with previous and future networks. NOMA-focused research initiatives will also examine novel use cases and applications, widening the potential impact of the technology in diverse fields.

NOMA will have a central role in 5G and beyond networks in achieving the objective of provision of high capacity, high spectral efficiency, low latency, and rich user experience such as guaranteed QoS. In that manner, by facing the challenges and seizing the potential of NOMA, low-cost and scalable communication systems in the post-5G era can support the huge number of connections expected in future wireless networks. Moving forward, their NP will play a key role in the research development, standardization, and technological evolution to continue steering NOMA toward a promising future for wireless communication.

3 INTEGRATION OF DEEP LEARNING WITH NOMA

The incorporation of DL with NOMA systems creates prospects for the enhancement of different facets of wireless communication. Since DL could approximate any functions that are amenable to the given data pattern, the conditioning of these patterns and relations could be employed to elevate the performance besides the efficiency of NOMA.

3.1 How deep learning enhances NOMA

3.1.1 Improved channel estimation

The channel estimation for NOMA is vital to the effective execution of SIC and power allocation techniques, hence the significance of the study. Some of the traditional methods that may be utilized for channel estimation may not do well, particularly in situations where the channel conditions are known to change very frequently. Specific methodologies, including DL models like convolutional neural networks (CNNs) and recurrent neural networks (RNNs), can considerably enhance the accuracy of channel estimation by learning the channel state from prior information and predicting the channel state in the future effectively. Compared with conventional models, these models can fit the more variable and cross channel and also offer more correct CSI for NOMA systems. For example, CNNs can feed large-scale channel matrices, extracting the right features for improving estimation while RNNs can adapt to temporal changes in the channel conditions and propose real-time adaptations and forecasts.

3.1.2 Dynamic resource allocation

Due to the complexity of the resource distribution, it is crucial to apply effective resource management in NOMA systems. These are the power control, the available bandwidth distribution, and the preferential schedule given to each user. Another important deficiency of DL algorithm is their ability to make accurate resource allocation decisions in real time by sensing dynamic changes in the network environment, as well as in the users' demand. The deep reinforcement learning (DRL) algorithms that may be used to design NOMA systems include the DQNs and the PPO. These algorithms can adapt to given situations and come up with reasonable decisions to utilize resources in the network properly to reach maximum efficiency. These features give the DRL agents the feedback from the network to continuously vary the power levels, the frequency bands, and the time slots such that the scarce resources are optimally adjusted and utilized to satisfy the users' demands.

3.1.3 Interference management

Interference management is one of the biggest obstacles that need to be overcome while designing NOMA systems. These sophisticated techniques in ML allow for interferer patterns to be recognized and subsequently, the power control and beamforming processes to improve on the interference. It will be seen that by accurately modeling the interference environment DL can cater for a reliable communication of all users of NOMA systems.

Due to GANs' ability to perform simulation and forecasting analysis, they can be harnessed to create experiences based on interference to design interference prevention measures. It can produce physically feasible interference patterns to enable network operators to design efficient mechanisms for controlling and mitigating interference in NOMA systems. Furthermore, with DL the models used for signaling can be trained to constantly learn and adapt to these variations of the interference while maintaining the quality of the communication.

3.1.4 User clustering

In NOMA systems, many times the users are grouped according to the conditions of the channels so that the resources can be effectively allocated. DL algorithms can change the dynamics of the cluster in real time based on mobility patterns, channel conditions, traffic loads, and many other scenarios. It also involves dynamic clustering to properly require the resources reducing the enhancement of total NOMA systems. It is also noteworthy that clustering algorithms can be better improved with DL approaches, including k-means and hierarchical clustering, to enable improved and dynamic user clustering. These algorithms can become wiser, and therefore function optimally in terms of the user clusters because they learn from the network environment constantly. Based on DL for clustering, one can also address more complicated user distribution patterns more effectively, which indeed contributes to supporting multiple types of users and improves the global performance of the network.

3.2 Key deep learning models used in NOMA

3.2.1 Convolutional neural networks

CNN is used when extracting spatial features in the data because CNNs work efficiently with data that is in a structured format such as CSI matrices. CNNs include convolutional layers, pooling layers, and fully connected layers, thereby making it possible for the appearance of comprehensive spatial structures. For the applications of CNNs in NOMA systems, they can be employed in channel estimation, user detection, and also interference prediction. Therefore, by training from the spatial data, the CNNs enhance their information delivery capacity with high accuracy to optimize the performance of NOMA systems. For instance, CNNs can be taught to identify specific patterns in the channel state that suggest the unsuitability of data transmission, thereby improving data throughput amid disruption.

3.2.2 Recurrent neural networks

RNNs are especially useful for handling sequential information, which is useful in data communications since the networks are implemented for time series. They are recurrent neural networks and they have connections that feature directed cycles, which means that they can keep information about previous inputs in the sequence. As a result, in the case of NOMA systems, the RNNs can be employed in predicting the mobility of the users, changes in channels, and traffic for the network. Other types of RNNs are LSTM layouts and GRUs layouts, which help in handling the vanishing gradient problem to offer a more accurate predictive model for sequential data. Through the use of RNNs, NOMA systems can accurately predict a user's behavior and set up a mechanism that can adjust the usage of resources accordingly to avoid disruptions of services while at the same time enhancing the QoS being offered.

3.2.3 Generative Adversarial Networks

Generative adversarial networks (GANs) are formed by two main models, the generator and the discriminator, both of which come together to produce realistic samples. Because, through data augmentation and simulation, GANs facilitate the performance of communication systems. In NOMA systems, GANs can be applied to provide realistic scenario interference, create synthetic data for the channel, and estimate user behavior. It can also help the network operators enhance their NOMA systems in terms of design and implementation to have better performance and reliability. First, GANs can help in developing a training dataset for DL when the original dataset is either sparse or prohibitively expensive while second, they can speed up the implementation of new and enhanced NOMA solutions.

3.2.4 Deep Reinforcement Learning

DRL is a reinforcement learning algorithm that is based on DL; thus, it is fitting for decision-making such as resource management and power regulation of NOMA systems. In DRL, an agent incorporates an action selection mechanism that lets him or her make a decision depending on the action-less environment and provides an understanding of the outcome in terms of reward or punishment. As for future research, this study concludes that the DRL algorithms in NOMA systems can improve resource management, power control, and user scheduling through realistic network conditions and user behavior. This enables the NOMA systems to meet and balance the requirements within the varying environment to make it work at its best. DRL can also enhance the scalable aspects of NOMA systems because it comprehensively tracks resources to achieve efficiency sufficient for the growing size and complexity of networks, as well as the changes in conditions.

3.3 Techniques for integrating deep learning in NOMA systems

The combination of DL with NOMA systems involves several techniques to ensure that the benefits of both technologies are fully realized. These techniques include:

3.3.1 Supervised learning for user clustering

Classification algorithms of supervised learning can then be utilized to categorize users according to the labeled data indicating historical channel conditions and traffic demands. Through applying DL to this data, NOMA systems will be able to realize user clustering simultaneously in real time meaning that resources will be allocated efficiently to give the systems a push and make them perform better. With supervised learning methods, the algorithm schemes can also be trained to perform effective updates under emerging user behaviors and network status at different periods constantly [2].

3.3.2 Unsupervised learning for resource optimization

The unsupervised learning algorithms including autoencoders and clustering can be used in the context of optimizing the resource in NOMA systems. These algorithms are also able to learn relationships in the data in the absence of labels, and the insights gained may be useful in determining power, bandwidth, and user assignment. It can also analyze the data that contains the hidden structures that are beneficial for better management of organizational resources.

3.3.3 Hybrid models for enhanced performance

A combination of multiple DL methods can be implemented to improve the performance of NOMA systems. For instance, CNNs and RNNs could be applied for channel estimation as well as users' mobility forecasts; GANs might be applied for interference emulation and obviation. These models can combine and solve more than one problem at a time and will guarantee efficiency in the NOMA system. The benefits of studying different types of DL techniques, evaluating them as a part, and combining them with the advantages of other techniques make the hybrid model a better solution for the issues represented by the NOMA systems.

3.3.4 Transfer learning for adaptability

Transfer learning models can be applied to changing pre-trained DL models and applying them in new settings and contexts. This means that advanced knowledge of DL systems is not necessary to develop sophisticated NOMA systems that will fit the network operators' requirements as the models can be fine-tuned to work with their specific NOMA data. Through this approach, one saves significant time on extensive training, and the implementation of robust new NOMA solutions is enhanced. Transfer learning also enables NOMA systems to reduce the amount of knowledge and experts needed, which in turn helps save on the time taken when learning and adapting to new technologies or networks.

4 METHODOLOGIES AND ALGORITHMS

4.1 Deep learning algorithms for NOMA

4.1.1 Convolutional neural networks for channel estimation

CNNs are specifically excellent in handling grid-like data like images and are thus suitable for CSI matrices processing in NOMA systems. CNNs derive spatial characteristics from CSI matrices and help the system comprehend channel scenarios better. The key concept of NOMA is that many users access the same frequency band, and signals from these users can intermodulate. Power control is very critical in increasing throughput rate where channel estimates need to be precise to have the right power control for the OFDM symbols. In this context, CNNs can be employed to obtain better results for predicting the channels' states, as the system can learn more detailed patterns based on the CSI data. This improved channel estimation helps in:

a. Optimizing resource allocation: These include the power and bandwidth control that can be achieved with a perfect channel estimated, which aims to support many users, hence improving the throughput.

b. Enhancing signal quality: This information is useful for the system to modify its transmitted parameters in a bid to be less disruptive of the channel and in the process improve the quality of the received signals.

c. Improving user experience: Through efficient and focused resource provisioning and improved signal quality, the users avail better data throughput rates as well as better link quality [3].

4.1.2 Recurrent neural networks for user mobility prediction

RNNs can also address the problem of sequential data; thus, they are considered suitable for predicting the mobility of users in NOMA systems. In a mobile communication environment, the users being dynamic and their positions about the base station as well as the channel conditions change over time, thereby affecting the resources required. This is because, as opposed to feed-forward networks, RNNs can remember earlier input through their so-called hidden state which makes it possible for them to capture temporal dependencies in the user movement pattern. By predicting where users are likely to move, RNNs help in:

a. Dynamic resource allocation: The ability to predict the movements of the users allows the system to be flexible in terms of resource provision and acquisition with a view of ensuring that the connectivity, as well as performance characteristics, is kept optimal.

b. Reducing handover failures: This way, by predicting the trajectory of the user and the utilization of the cells, the handover process itself can be done more efficiently and the connections lost are minimized.

c. Optimizing network performance: Mobile user movements also facilitate the more efficient organization of network resources where demand is high toward guaranteeing optimum coverage and bandwidth.

4.1.3 Generative adversarial networks for interference simulation

Automotive GANs are two or more neural networks, a generator and a discriminator, that generate realistic data. In NOMA systems, GANs can play the role of interference emulation, as this model can generate several channel data that reflect real-life conditions. With regards to interference simulations, one can essentially never simulate interference enough to get the right results in comparisons done for communication systems. Other advantages of GANs include:

a. Building diverse training data: Many interference scenarios can be achieved by GANs, which supply other DL models with diverse training data that would enhance their generality and robustness.

b. Enhancing interference management: Interference data that is simulated is useful in the formulation of interference management methodologies that are to be used in the system in question or in solving interferences that are likely to be encountered in the system in question.

c. Improving system reliability: Given that models are trained to different types of interference, the use of the system is shielded from potential interference episodes, resulting in enhanced performance and reliability.

4.1.4 Deep Reinforcement Learning for resource allocation

DRL integrates DL and reinforcement learning, which means it can be utilized as the decision-making algorithm in the NOMA systems including resource allocation and power control. In DRL, an agent acts as a learner, where it chooses its actions based on the interactions with the environment, and it gets feedback in terms of prices. Thus, DRL is a suitable component for NOMA systems, as it follows the changes in the network characteristics and maximizes the resource allocation in real time. Benefits of using DRL include:

a. Adaptive resource management: Rather than demanding predefined strategies DRL agents can learn about the bad and the good parts of available network resources by analyzing the current and future user demand and the channel conditions of the network environment.

b. Improved efficiency: By applying the concept of DRL, resource allocation, and utilization in real time, it is possible to improve performance all through the network and make sure resources are utilized in areas where they are most wanted.

c. Scalability: The combinations and characteristics of the DRL algorithms demonstrate their ability to manage large, complicated networks, and thus, they are advisable for future communication networks with high user density and different demands.

4.2 System models for NOMA

4.2.1 NOMA system model

In a normal NOMA system, several users communicate simultaneously using the same frequency channel. The main issue in NOMA is inter-user interference which should be well mitigated while aiming at achieving the highest possible system capacity and at the same time providing reasonable fairness. The relevance of a robust NOMA system model includes:

a. Resource utilization: Since operators have an accurate model of the network, they are better placed in terms of power usage and scheduling which leads to better network utilization.

b. Performance analysis: The detailed system model enables high-level performance evaluation under given conditions and the potential changes and enhancements to be considered.

c. Design and testing: The creation of the system model is mandatory for the development of new algorithms and techniques and for testing them in an environment that can be closely monitored and controlled before being implemented.

4.2.2 Power allocation in NOMA

Allocation of power in NOMA systems is very sensitive since it has a direct effect on the system efficiency in addition to user satisfaction. The relevance of optimizing power allocation includes:

a. Throughputs: When power is allocated optimally, the total system throughput is enhanced and delivers higher data rates to the users.

b. Ensuring fairness: Correct power control should maintain balance in using the network's resources so that all users are treated equally even if their channels are bad.

c. Reducing interference: Through selective use of power control, the system can reduce the infringement with other users and enhance the quality and stability of the signal.

4.2.3 Interference management

Interference management is an important aspect that should be considered in designing NOMA systems to ensure high-quality communications. Effective interference management strategies are relevant for:

a. Improving signal acquisition: Reducing interference causes increased SNR, which means the signal is more easily recovered.

b. Increasing capacity: This way, the system can thus accommodate more users in the same bandwidth which leads to efficient expansion of the total network.

c. Improving user experience: Interference directs degradation of the general quality of communication by reducing the number of available connections in addition to the raw data rate, ultimately depriving users of a satisfying communication experience.

4.2.4 User clustering in NOMA

In clustering the users, one has to put them in classes according to the conditions in the available communication channels and their requirements. The relevance of user clustering in NOMA systems includes:

a. Summarizing results and conclusions: Optimizing the usage of the allocated radio resources, and the users possessing similar channel characteristics are grouped to achieve the highest system result.

b. Reducing complexity: In essence, the management of users in clusters enables the system to minimize the complexity of resource allocation along with the management of interferences, thus making the system more efficient to apply and run.

c. Enhancing fairness: It also helps to achieve a better usage of the network resources and thus makes the interference measurably more fairer among users [4].

5 RESEARCH PAPERS AND THEIR CONTRIBUTIONS

Modern wireless communication benefits a lot from a technology called NOMA, which makes user connectivity and spectral efficiency better. One method of improving the system may involve using DL into it because this could enhance various performance measures, e.g., channel estimation, signal detection as well as power optimization. The most recent progress in the area is summarized in this literature review.

a. Signal identification in NOMA wireless systems: This research is in the early stage and applied a DL-based technique to identify signals in NOMA systems. It exhibits that neural networks were feasible for distinguishing multiuser signals, ensuring better detection accuracy and system performance [5].

b. Deep learning for multiuser MIMO systems: This work will consider the application of DL to channel estimation and power optimization in multiuser MIMO NOMA systems. The proposed algorithms have significantly increased the system's capability in interference management and resource allocation optimization [6].

c. Intelligent reflecting surface-assisted beamforming–NOMA networks: In this paper, the authors discuss intelligent reflecting surfaces correlated with beamforming in NOMA systems. It thus highlights that DL is used to optimize IRS configuration for good signal quality with less power consumption [7].

d. Performance analysis of wireless-powered cognitive IoT networks: The work in this research paper investigates the performance of wireless-powered cognitive IoT networks using NOMA and proposes a DL strategy for power allocation to optimize network throughput [8].

e. Deep learning for signal detection in NOMA systems: The signal detection-oriented research emphasizes the issue solved by DL and illuminates how DL has been used to enhance detection accuracy in NOMA systems. Its findings are that DL significantly reduces error rates and also contributes to improving the robustness of signal detection [9].

f. Machine learning for next-generation multiple access: This chapter describes the more general impact of ML, including DL, in the context of next-generation multiple access technologies. In this context, different state-of-the-art advances and future directions that become important for integrating AI for advanced wireless communications using NOMA are covered [10].

g. Electronics sensors application: It delves into DL for channel estimation and power optimization in NOMA systems. This illustrates DL for enhancing the system's performance under practical conditions [6].

h. Detection in wireless systems using DL-based technique: This chapter is a comprehensive study on the application of DL in the field of signal detection in wireless systems, with particular regard to problems and their solutions in NOMA frameworks [11].

i. IEEE conference tutorials: Everything concerning the tutorials to the recent advances of NOMA, with a particular interest in the role embedded DL plays in realizing further system efficiency and enhancing user experience [12].

j. Toward next-generation NOMA systems: This paper presents an overview of a few DL-based techniques developed recently toward the realization of next-generation NOMA systems, showcasing one more time the pivotal role that AI techniques play in complex wireless communication environments [13]. Among other recent upcoming wireless communications, integration of DL within NOMA systems is important. The reviewed papers show considerable growth in signal detection, channel estimation, and power optimization, which lead to the beneficial high-performance outcomes of NOMA.

6 CHALLENGES AND FUTURE DIRECTIONS

6.1 Current challenges in integrating deep learning with NOMA

6.1.1 Computational complexity

Another primary concern that can be attributed to the integration between DL and NOMA systems is that of computational complexity. Training DNNs need considerable computational power like GPUs or some specialized hardware engines. This complexity manifests in several ways:

a. Training time: As for the DL models, when the training dataset is large, only the models' training process will take a lot of time and consume a lot of resources. This is even more restrictive for certain applications such as real-time applications in NOMA architecture which require frequent model updates. Learning takes several cycles and back-propagation steps to update weights and while doing this can be complex. Due to the nature of wireless environments, the patterns of traffic may change over time and this translates to the need to frequently retrain the models which increases the challenge.

b. Inference latency: In deeper structures, additional layers create more complicated models, which may lead to inference latency. This is very disadvantageous for applications that need a positive response to a situation or event without much delay. Inertia can prospectively ruin a real-time application such as autonomous vehicles, augmented reality, or industrial processes that require split-second decision-making.

6.1.2 Data requirements

Data labeling is crucial for achieving high accuracy in DL models due to the requirement of huge datasets. In the context of NOMA systems, acquiring such data poses significant challenges:

a. Data collection: Gaining adequate training data, which would necessarily contain subsets of the varied conditions of NOMA systems, is a cumbersome process. This data shall incorporate several forms of users' mobility, channels' state, and possible interferences. The need to gather data results in the conduct of huge measurement campaigns within wireless environments which can be very cumbersome and costly. Lastly, data collection needs to reflect at least the variations in users, devices, and network topologies to be comprehensive.

b. Data labeling: In supervised learning problems often labeling of the data requires a lot of human input and can be time-consuming as well as involves errors. Self-generated label assignment techniques need to be efficient to address the challenges of wireless scenarios. Manual labeling may be subjective and may include biases that are not inherent in automated methods which require model/species tagging algorithms to properly tag data in dynamic and heterogeneous scenarios.

6.2 Future trends and innovations in NOMA with deep learning

6.2.1 Federated learning

Federated learning offers a promising approach to address data privacy and security concerns: As for the training process, federated learning allows the models to be trained on devices without sending data to central servers to be trained on a central server, improving privacy. Using such an approach minimizes the need to collect large amounts of data, including personal identification information, centralizing them for potential hacking or breaching, and violates data protection policies in many countries. Model aggregation: This approach has the advantage of updating a model with data from different devices, potentially creating more general and finalized models. To facilitate the quick formulation of the model updates, protocols like federated averaging (FedAvg) are employed to prevent the global model from not being updated by the collective learnings of all the devices.

6.2.2 Edge computing

Leveraging edge computing resources can significantly enhance real-time processing capabilities:

a. Increased efficiency: Implementing DL models nearer the user through the network reduces latency and latency times. This is because edge computing allows for real-time data processing, thereby giving great effectiveness to those applications that are sensitive to latency.

b. Real-time adaptation: Edge computing helps in real-time model updates to many changes in the network environment, resulting in improved functioning of NOMA systems. Since most data processing takes place at the edge of networks, NOMA can enhance responsiveness in concurrently changing conditions while maintaining QoS.

7 CONCLUSION

This chapter provided a comprehensive overview of NOMA systems and their potential to be revolutionized by DL integration. NOMA has promised huge benefits in network capacity, spectral efficiency, and latency for 5G and beyond. Their primary issues are in channel estimation, user clustering, interference management, and resource allocation, which can be effectively resolved using DL algorithms. Integrating DL techniques into the NOMA scheme will make its systems more efficient, agile, and user-centric. The future of NOMA lies in integration with AI-driven network optimization.

REFERENCES

1. Al-Imari, M., Xiao, P., Imran, M. A. and Tafazolli, R. (2014). Uplink Non-Orthogonal Multiple Access for 5G Wireless Networks, in 2014 *11th International Symposium on Wireless Communications Systems (ISWCS)* Barcelona, Spain, pp. 781–785, doi:10.1109/ISWCS.2014.6933459.
2. Cui, J., Ding, Z., Fan, P. and Al-Dhahir, N. (2018, November). Unsupervised Machine Learning-Based User Clustering in Millimeter-Wave-NOMA Systems. *IEEE Transactions on Wireless Communications*, vol. 17, no. 11 pp. 7425–7440, doi:10.1109/TWC.2018.2867180.
3. Anu, S. and Raveendran, T. (2021). CNN based Channel Estimation Using NOMA for mmWave Massive MIMO System. In *2023 IEEE Statistical Signal Processing Workshop (SSP)* (pp. 349–353). IEEE.

4. AbdelMoniem, M., Gasser, S. M., El-Mahallawy, M. S., Fakhr, M. W. and Soliman, A. (2019). Enhanced NOMA System Using Adaptive Coding and Modulation Based on LSTM Neural Network Channel Estimation. *Applied Sciences*, vol. 9, no. 15, p. 3022, doi:10.3390/app9153022
5. Signal Identification in Non-Orthogonal Multiple Access Wireless Systems. (2023, September 11). *Propulsion Tech Journal* (https://www.propulsion-techjournal.com/index.php/journal/article/view/697)
6. Deep Learning-Based Detection Algorithm for the Multi-User MIMO. (2023). *Electronics*, vol. 13, no. 2, p. 255, doi:10.3390/electronics13020255
7. Intelligent Reflecting Surface-Assisted Beamforming-NOMA Networks. (2023, August 21). *IET Communications*, doi:10.1049/cmu2.12667
8. Performance Analysis of Wireless-Powered Cognitive IoT Networks. (2023, August 21). *IET Communications*, doi:10.1049/cmu2.12667
9. Thompson, J. (2019, August). Deep learning for signal detection in non-orthogonal multiple access wireless systems. In *2019 UK/China Emerging Technologies (UCET)* (pp. 1–4). IEEE. doi: 10.1109/UCET.2019.8881888.
10. Machine Learning for Next Generation Multiple Access. (2024). *IEEE Communications Society*. (https://ngma.committees.comsoc.org/events-2/)
11. Narengerile and John S. Thompson. (2019). Deep Learning for Signal Detection in Non-Orthogonal Multiple Access Wireless Systems, in *2019 UK/China Emerging Technologies (UCET)*, pp. 1–4.
12. Tutorials | IEEE International Conference on Communications. (2024). *IEEE ICC* (https://icc2024.ieee-icc.org/program/tutorials)
13. Dai, L., Wang, B., Yuan, Y., Han, S., Chih-Lin, I., & Wang, Z. (2015). Non-orthogonal multiple access for 5G: solutions, challenges, opportunities, and future research trends. *IEEE Communications Magazine*, vol. 53, no. 9, pp. 74–81.

CHAPTER 7

Advancements in machine learning techniques for optimization of massive MIMO design

Avula Mahathi, C. Kishor Kumar Reddy,
Kari Lippert and Marlia M. Hanafiah

1 INTRODUCTION

Wireless communication technology has evolved over the last three decades due to advanced research and development needs for higher data rates, better reliability, and greater coverage. One of the main technologies that have been developed to satisfy these requirements is multiple input, multiple output (MIMO), also known as spatial division multiplex [1]. MIMO technology innovatively transposes wireless communication systems to utilize multiple antennas at the transmitter and recipient sides. Such multi-antenna solutions increase the transmission capacity and the probability of successful communication in wireless systems, and therefore the efficiency and reliability of these systems.

DOI: 10.1201/9781003514336-7

In this chapter, it is important to dig into the concepts of MIMO technology, how it is being implemented, and the ways in which machine learning (ML) takes part in actualizing MIMO design. We will delve into the three core principles of MIMO: namely, DPs involve spatial diversity, spatial multiplexing, and beamforming. These principles are central to the operation of MIMO technology, ensuring that the data transmitted is more efficient and coherent than in a system using a single antenna.

1.1 Background to multiple-input multiple-output technology

A new technical innovation in mobile communication can be illustrated by MIMO. It is where the basic communication platform employs a number of antennas at the base as well as at the terminals to enhance communication performance. MIMO can be defined as a technology that utilizes the frequency correspondence of the space in the wireless channel to achieve multiple simultaneous signal pathways. This simultaneous transmission increases the capacity as well as reliability of wireless communication; it aims to meet this ever-increasing demand for high-speed data transfer.

The overall significance of MIMO technology is that it offers solutions to enhance wireless communication in several possible ways [2]. MIMO systems are capable of transmitting more data streams, and that in turn leads to increasing the data rate of the wireless communication system, which is very beneficial where large bandwidth is required, such as video services, interactive games, and big data services. Another opportunity presented by MIMO is spatial diversity, which implies the use of multiple paths in the transmission of data. This assists in reducing the impact of fading and interference, hence improving the probability of the transmission link. Spatial diversity enables the system to maintain good communication links even if a number of paths are fading [3]. Moreover, MIMO offers enhanced spatial resource utilization since it endorses high-order spatial multiplexing by applying superior signal processing techniques to isolate numerous data streams within the same frequency range, besides optimally exploiting the radio band. In addition, a technique applied to MIMO systems, namely beamforming, concentrates the transmission power toward certain directions rather than covering all directions to an equal extent. For the V2X system, this targeted approach increases the strength and coverage of signals, which is crucial to successful transmission, particularly in areas with high interference, such as urban environments.

MIMO technology is utilized in various applications and technologies, significantly impacting the wireless communication landscape.

Wi-Fi Networks: MIMO is the base of most of the new Wi-Fi standards, like 802.11n (Wi-Fi 4), 802.11n, 11ac (Wi-Fi 5), and 802.11ax (Wi-Fi 6). The capabilities of these standards incorporate MIMO as a means to offer higher data rates and performance, thus enhancing the operation of wireless internet networks.

4G LTE: Some of the current wireless 4G networks, known as long-term evolution (LTE), use MIMO technology to increase data rates and system spectral efficiency. MIMO-assisted LTE networks offer high speeds to allow users the kind of internet experience expected when on the move.

Fifth generation (5G) Networks: Massive MIMO (mMIMO), which uses a lot of antennas, is a major component of 5G, the next generation of wireless technology. To achieve ultrahigh data rates, low latency, and vast connection needed for 5G applications like the Internet of Things (IoT), driverless cars, and smart cities, mMIMO is a must. For 5G networks, frequency ranges up to 300 GHz have been investigated [4].

The development of MIMO from its inception to its present application in sophisticated wireless networks emphasizes how important a role it will play in determining the direction of wireless communication. MIMO technology provides more effective, dependable, and high-capacity wireless communication systems to fulfill the demands of an increasingly interconnected world by overcoming the drawbacks of conventional single-antenna systems. Whereas MIMO systems use several antennas at both ends, single input, single output (SISO) systems use a single antenna at both the transmitter and the receiver. Because MIMO channels have advantages over SISO channels in terms of multiplexing, diversity, and array gains, MIMO systems have increased capacity and dependability [5].

Table 7.1 offers a detailed and enhanced comparison of different MIMO technologies that were used in wireless communications networks. Starting with SISO, which uses an individual antenna both for transmission and reception, it is quite simple, yet it has very low capabilities and is suitable for basic wireless communication needs only. Transitioning to single-input-multiple-output (SIMO), this configuration improves the signals' reception by using one transmitting antenna and multiple receiving antennas; it is efficient in situations where signal shielding is crucial. Multiple-input-single-output (MISO) technology enhances the transmission diversity as the system deploys multiple antennas for transmitting and one for receiving, optimizing the uplink and the receiving link. Last in

TABLE 7.1 Comparison of MIMO technologies

Technology	Number of Antennas	Key Features	Applications
SISO	• Single input, single output	• Simplest form, limited capacity	• Basic wireless communication
• SIMO	• Single input, multiple output	• Improved reception reliability	• Early Wi-Fi standards
• MISO	• Multiple input, single output	• Enhanced transmission diversity	• Improved uplink performance
• MIMO	• Multiple input, multiple output	• Increased capacity, spatial multiplexing	• Wi-Fi, 4G, 5G, LTE, high-bandwidth environments

the list of antenna choices is MIMO, which involves multiple transmitting and receiving antennas; they enhance data rate and spectrum efficiency through spatial multiplexing and diversity, as in the latest wireless standards such as Wi-Fi, 4G, and 5G, which are most useful where high data rate and reliability are needed.

1.2 Role of machine learning in optimizing massive MIMO

Technological advancements in wireless communication make the operation of MIMO technology complicated. Since MIMO systems deploy more antennas and aim for advanced signal processing strategies, optimizing these systems gets more complicated. Although heuristic algorithms and manual tuning have been applied as standard solutions for improving MIMO performance in the past, they can only provide a general solution and are short of the flexibility needed in the growing complexity and variation of wireless communication systems. In the context of mMIMO systems, recent research has found the application of ML techniques highly effective. One of the advantages of ML algorithms is that they do not require human intervention; rather, relying on large datasets and computational resources, they are capable of learning patterns and relationships that underlie the process of optimization. These techniques facilitate the ability of MIMO systems to evolve with different channel links, reduce interference, and optimize channel capacity without any a priori formulations determining how the system should operate.

Another advantage of ML in the design of MIMO is the ability to maximize the large number and to manage the high level of complexity and nonlinearity of modern communication systems. Conventional optimization

TABLE 7.2 Increasing complexity of MIMO systems and the need for ML techniques

Year	Complexity of MIMO Systems	Need for ML Techniques
2000	• Low	• Low
• 2005	• Moderate	• Low
• 2010	• High	• Moderate
2015	• Very high	• High
• 2020	• Extremely high	• Very high
• 2025	• Exceptionally high	• Extremely high

techniques are generally static in modeling the electromagnetic interactions between multiple antennas, users, and other environmental factors. On the other hand, pure ML algorithms can process big data and generate useful insights in parallel to achieving higher MIMO system performance rates, which will only get better with time. As evident from the following graph, MIMO systems have become increasingly complex over a number of years as hardware advances and the need for converged high-rate and reliability solutions escalate. Similarly, the application of ML techniques has increased as the current optimization procedure demands enhanced identification methodologies to solve the problem of rising wireless networks. The shedding of new light on the understanding of mMIMO through the use of ML will allow researchers and engineers to fully explore the potential of this new technology to develop the next big wireless communication systems.

The MIMO system complexity is shown in Table 7.2 as it increases over time, from a low level in 2000 to an unusually high level by 2025. In line with this, the demand for ML approaches grows; it was little in 2000 but will progressively increase by 2025 to meet the demands of extremely complex MIMO systems.

2 FUNDAMENTALS OF MIMO TECHNOLOGY

MIMO is a revolutionary technology in wireless communication systems, which actually incorporates multiple antennas at both the transmitter and the receiver ends. This makes it possible to use multiple frequency channels for both transmission and reception of data, thus improving the quality, capability, and efficiency of wireless transmitting networks. Thus, originally, MIMO technology takes advantage of the spatial aspect of the wireless communication channel by employing the spatial diversity incorporated in multiple-antenna systems. MIMO systems operate by sending signals over at least two distinc spatial paths, and this makes it possible to avoid some of the difficulties experienced when using a single antenna, namely fading and interference.

In addition, there is the MIMO technology which makes use of space diversity to allow spatial multiplexing, a process that allows several streams of independent data to be transmitted and to be received independently over the same frequency band. This leads onto impaired data throughput, increased data rates, and also efficient use of the wireless network spectrum.

In conclusion, MIMO technology can be considered as being a major improvement in the current wireless communication systems, as it increases the number of transmitting and receiving antennas, making the systems more reliable and bringing new improved spectral efficiency. This shows how it has been adopted in different wireless standards such as Wi-Fi, LTE, and 5G to show its relevance in determining how the future wireless network should be.

2.1 Spatial diversity, multiplexing and beamforming techniques

Spatial diversity is also another aspect that was added by MIMO system since it can take advantage of the diversity of space through its multiple antennas. This is due to the fact that spatial diversity sends out the same signal through various steep paths in space which decreases fading and interference, thereby increasing the signal dependability. While spatial division helps to achieve multiples of individual links using multiple antennas for transmission and reception, spatial multiplexing facilitates MIMO systems to send multiple independent signals in parallel in the same frequency. This is done by using the space dimension to send several data streams through different antennas, thus increasing the data rate or data throughput of the whole system.

Together, these two techniques are indispensable constituents of the expertise of wireless communication since they support the propagation of signals with greater reliability and the flow of data with higher throughput in a system of spatial diversification and multiplexing.

Beamforming is a transmission power management method of the MIMO in which signals are steered in specific directions [6]. Notably, beamforming revolves around the act of helping the transmission beam direct its focus toward the target receiver while limiting the inference from other directions, hence improving signal quality and transmission efficiency. Beamforming can be effectively applied in MIMO systems by applying either analog or digital beamforming methodologies. The first one is used at the antenna level where phase and amplitude of the signal are modified analogously to achieve beamforming, while the second operates on a certain set of algorithms to form the beam. In general, it should be noted that the use of beamforming techniques is crucial in order to enhance the performance characteristics of MIMO systems such as reliability, coverage, and pilot resources in current wireless communication technologies.

Table 7.3 provides a comparative analysis of three key techniques used in MIMO systems: spatial diversity, spatial multiplexing, and beamforming. The diversity would be spread through space, and the signals would be multiplexed. Spatial diversity involves the use of two or more antennas to send the signal through different spatial paths, thus increasing the reliability of the signal by combating fading and improving the signal-to-noise ratio. This technique is widely used in areas with barriers or fading signals, which makes it important in ensuring uninterrupted transmission.

Spatial multiplexing allows for transmission of multiple data streams through multiple antennas, hence enhancing the amount of data that

TABLE 7.3 Comparison of spatial diversity, multiplexing, and beamforming techniques in MIMO systems

Technique	Description	Key Features	Applications
Spatial diversity	• Utilizes multiple antennas to transmit the same signal through different spatial paths to improve reliability.	• Enhances signal reliability by mitigating fading and improving signal-to-noise ratio.	• Used in environments with obstacles or signal fading.
• Spatial multiplexing	• Simultaneously transmits multiple data streams using multiple antennas to increase data throughput.	• Boosts data throughput by exploiting spatial dimensions, enabling parallel transmission of data streams.	• Commonly used in high-bandwidth applications like 5G.
• Beamforming	• Focuses the transmission or reception of signals in specific directions, improving signal strength and quality.	• Directs signals toward the intended receiver, reducing interference and enhancing signal coverage.	• Widely employed in wireless communication standards to improve range and reliability.

can be transmitted. Spatial multiplexing exercises different transmit and receive dimensions to allow several data streams to be transmitted simultaneously, which is why it is used in high-throughput applications like 5G communications. They enhance the signal strength and quality by steering the transmission or reception of the signal in some specific directions normally referred to as beamforming. Compared to a more dispersed pattern that sends signals in multiple directions, beamforming focuses the signal in the direction of the intended receiver which increases coverage and signal quality. This technique is prevalent in wireless communication standards to enhance the coverage distance as well as the fundamental performance of a system.

3 ADVANCED TECHNIQUES IN MASSIVE MIMO DESIGN

The important study and knowledge that are brought together in "advanced techniques in mMIMO design" include a detailed analysis of the state-of-the-art techniques and approaches utilized in the mMIMO systems [7]. Since data rates, new spectral efficiency, and better performance of the network engineering demands are constantly on the rise, researchers and practitioners are embracing innovative solutions to accommodate these trends. This section tries to provide insight into the first version of MIMO technology and elaborate on the most advanced measures intended to enhance system performance, dimensionality, and efficiency.

From the maximum likelihood channel estimation process to the user pairing procedure and the power allocation scheme in the non-orthogonal multiple access (NOMA) system, each section in this chapter will provide insight on the complexities of the mMIMO system. Through introduction of scenarios and difficulties related to the implementation of such systems, the audience gets to know that there is more to MIMO systems than just the theory behind the mathematically derived large-scale systems. Thus, this chapter is an effort to provide the researchers, engineers, and industry professionals with the understanding and directions in the modern MIMO design by critically analyzing these methodologies to create an innovative wireless communication system.

3.1 Channel estimation

It is extremely important for the wireless channel associated with the transmitter and the receiver to be properly characterized in the case of mMIMO systems, through channel estimation. Because of the large number of antennas at both transmitter and receiver sides, any naive approach to channel estimation might become cumbersome. In the case of mMIMO, pilot sequences are usually employed to attempt channel response measurement at each antenna. However, several challenges have been identified out of which the problems of designing efficient pilot sequences and accurate estimation of the channel while suffering from pilot contamination issues are notable. One of the key indicators used to assess the effectiveness of wireless systems is capacity, which also functions as a practical manual for designing both the broadcast and received signal processing [8]. By reflecting or refracting incident signals in appropriate directions, the metasurface can optimize the wireless channel by enhancing signal strength and lowering interference [9].

3.2 MIMO precoding and signal detection

MIMO precoding aims to maximize the quality of the received signal at the receiver end by steering and tuning of the transmitting signal at the transmitter end. In mMIMO, the precoding is more challenging due to the large number of antennas together with the fact that the users are separated in both time and frequency domains. Amid interference control and signal quality enhancement, the application of precoding technology, encompassing ZF and MMSE, for example, is prominent. However, the signal detection techniques are concerned with the rebuilding of the transmitted symbols from the received signal. As demonstrated earlier, in a mMIMO system, the problem of detection becomes prominent due to the large dimensions of arrays; therefore, robust algorithms are needed to detect signals in the presence of noise and interference.

3.3 Hybrid beamforming

Digital and analog beamforming methods are combined in hybrid beamforming to achieve efficient beamforming while also reducing the need for additional hardware demodulation. However, in mMIMO scenarios, hybrid beamforming is adopted because full digital beamforming is impossible due to the large number of antennas. Hybrid beamforming simply splits beamforming into two sections, a digital baseband and an analog radio frequency, that offer a balance between performance and complexity. But, it is still an open research problem on what kind of hybrid beamforming architectures and algorithms should be employed so as to achieve great computational efficiency and performance robustness to the channel variations. The three objectives of mMIMO beamforming are transmission matrix formulation, transmitting antenna selection (AS), and accurate channel matrix estimation [10].

3.4 Distributed configurations

Distributed architectures in mMIMO systems are related to division of processing loads and control between the distributed antennas. This is especially pertinent in a situation where centralized processing is going to be complicated by the sheer number of antennas or the limitations on the communication. Integrated and distributed solutions make such systems further accommodating for large-scale MIMO operations by balancing resources and managing interference. However, the development of algorithms that allow for an orderly transmission and/or reception between distributed antennas and that will ensure acceptable performance of the overall system is challenging.

3.5 User pairing and power allocation in Non-Orthogonal Multiple Access

In NOMA, an important design principle involves optimizing either the sum-rate or individual rates through choosing the power coefficients and user pairing for users with similar channel conditions. When it comes to the large-scale systems like mMIMO, NOMA, conventional user pairing, and power allocation turn into more complex issues due to more number of users as well as the number of antennas. To reduce the interference between paired users and ensure effectiveness of data transmission, complex mathematical algorithms including convex optimization algorithms, game theory, and ML algorithms are applied to design the efficient user pairing and power allocation. However, the complexity of these algorithms and their dependence have made them very sensitive to channel uncertainties and feedback overhead which is very much really a problem in real-world implementations.

These subsections focus on important and essential features of the advanced techniques in the system of mMIMO where the practical implementations and related problem areas are briefly mentioned as well.

TABLE 7.4 Techniques and considerations in advanced massive MIMO design

Subsection	Description	Techniques and Methods	Challenges and Considerations
3.1 Channel estimation	• Techniques and challenges in estimating channel conditions for optimizing system performance.	• Estimation algorithms, pilot sequences	• Pilot contamination, channel modeling complexity
• 3.2 MIMO precoding and signal detection	• Advanced methods for signal processing and detection in massive MIMO systems.	• Precoding algorithms, detection schemes	• Complexity, computational overhead
• 3.3 Hybrid beamforming	• Utilization of hybrid beamforming techniques for optimizing signal transmission and reception.	• Analog and digital hybridization, codebook design	• Analog-digital interface, calibration issues
• 3.4 Distributed configurations	• Exploration of distributed MIMO configurations and their implications in massive MIMO design.	• Cooperative schemes, decentralized processing	• Synchronization overhead
• 3.5 User pairing and power allocation in non-orthogonal multiple access	• Strategies for pairing users and allocating power efficiently in non-orthogonal multiple access systems.	• Resource allocation algorithms, user grouping	• Interference management, fairness constraints

The techniques and difficulties involved in designing advanced mMIMOs are listed in Table 7.4. These include distributed configurations, hybrid beamforming, channel estimation, MIMO precoding and signal detection, user pairing, and power allocation in NOMA systems. Table 7.4 gives a brief overview, identifies the most important strategies and tactics used and addresses issues and concerns that must be resolved for successful use in real-world settings for each subsection.

4 SIGNAL DETECTION, HYBRID BEAMFORMING, AND MACHINE LEARNING APPLICATIONS

This section provides an understanding of how the application of ML techniques goes hand in hand with signal detection/hybrid beamforming for MIMO systems. On the other hand, signal detection is critical to almost every application of artificial intelligence (AI) and ML algorithms in that it is a primary process in which ML algorithms need to improve the performance of signal detection processes [11]. These algorithms are capable of smartly conducting analysis on different patterns of signals, can self-adjust itself with varying environmental conditions, and can also tune itself in terms of different detection limits in real time. In addition to this, the use of ML for deriving hybrid beamforming in MIMO systems allows for dynamic adaptations of the beamforming weights and phase so that maximum coverage for signals is achieved while minimizing interference. In particular, they requested that MIMO systems utilize ML techniques, so as to be able to efficiently operate in heterogeneous and dynamic environments and offer unprecedented levels of adaptability.

4.1 Power allocation, user pairing, and machine learning optimization

This subsection is aimed at presenting the effects of interacting ML in the design of power control and user association in MIMO. Serving with various channel conditions, traffic patterns, and quality of service requirements, ML algorithms enable MIMO systems to control the transmit power resource. These algorithms can then adapt from the past data recorded and the feedback obtained from the live environment in order to achieve the most desired power distribution to achieve the maximum available data rate and energy efficiency metrics all under rock-solid connectivity assurance. Also, the MAC layer, with the help of ML algorithms, enables intelligent pairing decisions based on signals and factors of the channel and the user mobility as well as the requirements of the applications. This means that MIMO systems have the opportunity to select and associate users to different clients in order to balance load, reduce interference, and consequently optimize the system's performance. This chapter well demonstrates how the integration of ML optimization opens up new opportunities for the development of MIMO systems in terms of functionality, versatility, and capacity in wireless communication networks.

5 APPLICATIONS AND CHALLENGES

As we delve into the diverse applications of mMIMO technology across different wireless networks, it's essential to recognize the underlying challenges and considerations that shape its deployment [12]. Understanding both the potential benefits and limitations is crucial for effectively harnessing the capabilities of mMIMO in real-world scenarios. In this section, we explore specific applications of mMIMO in 4G, 5G, and Wi-Fi networks, shedding light on how this technology revolutionizes wireless communication. Before delving into these examples, let's first examine the key challenges, particularly in terms of hardware complexity and scalability, that must be addressed to unlock the full potential of mMIMO.

5.1 Applications in 4G, 5G, and Wi-Fi networks

Specifically, mMIMO technology is applicable in 4G, 5G, and Wi-Fi wireless communication standards. mMIMO proves to be a very valuable tool in 4G LTE networks because it brings out a step change enhancement in both the spectral efficiency and the data rates as well as the capacities of networks. Compared with the 3G system, the 4G network makes use of a number of antennas at both transmitter and receiver ends to get improved coverage, minimized interference, and higher data rate, so that the user can have faster connection and reliable data service for heavy applications like video, voice, industry, and gaming. mMIMO has a crucial role to perform in achieving the abovementioned goals in 5G networks particularly by effectively facilitating ULR and mMTC. The field of wireless communication has experienced a profound metamorphosis, progressing through multiple technological generations that have brought about notable advancements in terms of functionality, effectiveness, and use. We are now on the verge of the 5G era, a new stage in wireless communication that offers previously unheard-of speeds, dependability, and connectedness, thanks to this evolutionary route [13]. Currently, commercial deployments of 5G telecommunications networks are underway. Cellular mMIMO systems are one of their key enabling technologies [14]. mMIMO, or the utilization of many antenna elements, is seen as a crucial technology that will enable the 5G and Beyond wireless ecosystem. On the physical channel of the wireless medium, the clever application of a myriad of antenna elements releases previously unheard-of flexibility and control. With mMIMO, the performance and operation of the physical wireless channel can be controlled or influenced by other network layers or wireless applications, opening up an entirely new set of parameters [15]. mMIMO in

5G networks due to its capability of massive connection, flexibility for the dynamic change of the channel and flexible system scale, offers extremely high performance in 5G networks to create opportunities for the applications which require high requirements for the connection such as automotive, smart city, and Industry 4.0. In addition, the solution of Wi-Fi network densification using the technology of mMIMO is currently being implemented to cater to the continuously rising need of higher wireless broadband access in highly built urban areas, public spaces, and business establishments. By using the spatial multiplexing and the beamforming properties supported by mMIMO, Wi-Fi networks that incorporate this technology offer higher data rates and coverage that can support various demanding applications, from smart homes to offices.

This section provides an insight into how mMIMO works in 4G, 5G, and Wi-Fi networks to deliver rich features and service benefits to users, enabling new and differentiated services for communication service providers and businesses.

5.2 Key challenges: hardware complexity and scalability

mMIMO has the potential to completely transform wireless communication, but a number of significant obstacles, mostly related to hardware complexity and scalability, are impeding its broad use. High data rate transmissions are possible thanks to mMIMO technology, which makes use of the HAPs' many antennas. Furthermore, by precoding, the massive number of antennas in an array can produce enough gain to overcome the free space path loss [16]. The complexity of the related hardware, such as digital signal processors, analog-to-digital converters, and radio frequency front-ends, rises with the number of antennas and radio frequency chains in mMIMO systems. Because the hardware must meet strict standards for power consumption, cost-effectiveness, and form factor while guaranteeing robust performance in real-world applications, this presents substantial design and implementation issues [17]. In addition, scalability in mMIMO installations is necessary to support an expanding number of connected devices and applications and meet the increasing demands of future wireless networks. Aspects like system architecture, antenna array design, signal processing methods, and network management protocols are all included in the concept of scalability. In order to address these issues, multidisciplinary research projects involving electrical engineering, signal processing, computer science, and telecommunications are needed.

The main goal of these projects is to create novel solutions that strike a balance between cost, usefulness, and performance. The obstacles related to hardware complexity and scalability in mMIMO systems are being significantly surmounted, nevertheless, by continuing research and development efforts. In order to fully realize mMIMO's promise and prepare the road for its smooth incorporation into next-generation wireless networks, researchers want to take advantage of developments in semiconductor technology, system integration, and algorithm design. A signal processing method called AS dramatically lowers the hardware needs of multi-antenna systems [18].

The performance gains in networks employing MIMO alone versus MIMO in conjunction with ML approaches are shown in Figure 7.1. Spectral efficiency, data rates, latency reduction, connection density, coverage extension, and throughput enhancement are among the performance measures across 4G, 5G, and Wi-Fi networks.

5.2.1 Key observations

Spectral efficiency and data rates: In 4G networks, MIMO alone (blue line) and MIMO + ML (red line) exhibit marginal gains.

Reduced latency: Using MIMO + ML in 5G greatly lowers latency as compared to using MIMO alone.

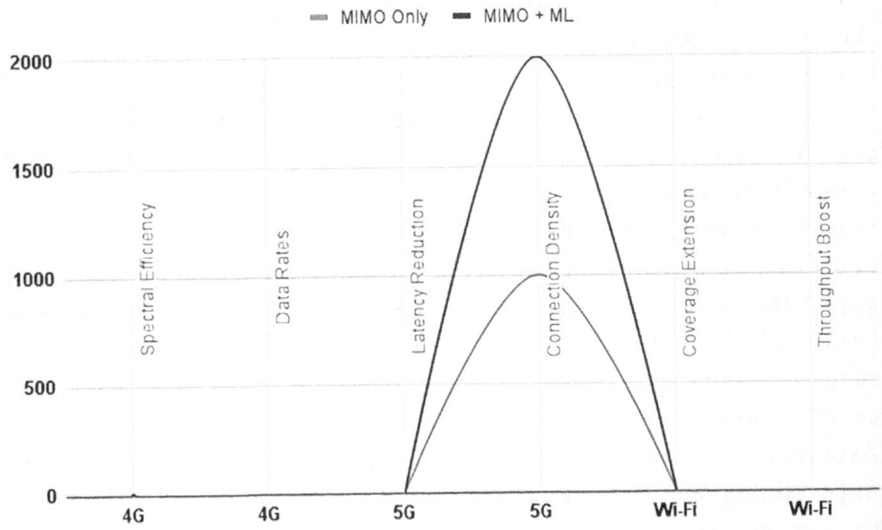

FIGURE 7.1 Performance improvements in networks due to MIMO and ML techniques.

Connection density: While MIMO only peaks at 1,000, MIMO + ML significantly increases connection density in 5G networks, peaking around 2,000.

Coverage extension: Using MIMO + ML improves coverage extension for Wi-Fi networks as compared to MIMO alone.

Throughput boost: When compared to MIMO alone, MIMO + ML approaches significantly increase throughput in Wi-Fi networks.

To sum up, the combination of ML with MIMO techniques yields significant improvements in performance for a variety of network technologies, particularly in 5G and Wi-Fi networks.

6 KEY FINDINGS AND SIGNIFICANCE

The primary contributions and conclusions of our chapter on advanced techniques in mMIMO design are outlined in this part. By investigating diverse approaches and obstacles, we have discovered a number of important realizations that enhance our comprehension and progress of mMIMO technology.

6.1 Key findings

Advanced techniques implementation: In this chapter, we explore the application of sophisticated methods to large MIMO systems, including user pairing, distributed configurations, hybrid beamforming, MIMO precoding, and channel estimation. We offer an understanding of the intricacies and difficulties related to every method, as well as possible fixes and optimization tactics.

Applications of machine learning: We emphasize how ML may be used to optimize mMIMO systems, especially in relation to signal recognition, user pairing, and power allocation. mMIMO design challenges can be effectively addressed by ML methods, which make use of substantial datasets and computer power.

Applications in 4G, 5G, and Wi-Fi networks: This chapter examines the various uses of mMIMO technology in 4G LTE, 5G, and Wi-Fi networks, among other wireless networks. We describe how mMIMO improves these networks' spectrum efficiency, data rates, coverage, and dependability, opening the door for cutting-edge wireless services.

6.2 Significance

Developing wireless communication: By addressing important issues and examining creative solutions in mMIMO design, the insights offered in this chapter help to progress wireless communication technology. Through technique optimization and ML, we improve wireless network performance, efficiency, and dependability.

Facilitating next-generation networks: The conclusions made here have a big impact on how next-generation networks—like 5G and Beyond—are developed. mMIMO technology holds the key to unlocking previously unattainable data rates, decreased latency, and greater connection density in future wireless networks by surmounting hardware complexity and scalability issues.

Encouraging innovation: The significance of ongoing research and innovation in the mMIMO field is emphasized in this chapter. mMIMO systems can be made even more capable and scalable by utilizing cutting-edge technologies like edge computing and quantum computing, along with embracing new trends in ML. This will spur innovation in wireless communication.

All things considered, the major conclusions and implications discussed in this chapter highlight the revolutionary potential of mMIMO technology in influencing the development of wireless communication networks in the future.

7 FUTURE DIRECTIONS

7.1 Emerging trends in machine learning for MIMO

Rapid advancements in ML are creating new avenues for enhancing MIMO systems. Modern ML applications for MIMO are set to transform wireless communication in a number of ways. The use of reinforcement learning (RL) techniques, which enables MIMO systems to adjust and maximize their performance in response to real-time environmental feedback, is one prominent trend. MIMO systems may learn and refine their decision-making processes thanks to RL algorithms, which increases spectral efficiency and facilitates more effective resource allocation. The potential of mMIMO antenna systems to achieve great spectrum efficiency in wireless communication systems is widely recognized. This is attributed to millimeter waves (mmWaves), which enable the deployment of numerous antennas at base stations (BSs) [19]. Using deep learning (DL) architectures in MIMO systems is another new trend. Recurrent neural networks (RNNs) and convolutional neural networks (CNNs) are two examples of DL models that provide effective tools for removing intricate patterns and correlations from MIMO channel data. MIMO systems can increase their accuracy in beamforming, channel estimation, and signal identification by utilizing DL approaches, which will ultimately improve system performance and reliability. Furthermore, federated learning (FL) is becoming more and more popular as a viable method for cooperative model training in dispersed MIMO networks. With FL, MIMO devices can work together to develop a global model while protecting the confidentiality and privacy of their data. MIMO systems can adjust to changing channel conditions and user mobility patterns because of this decentralized learning paradigm, which improves network resilience and scalability. Owing to the extremely quick expansion of wireless networks, the introduction of the sixth generation (6G) heralds the start of novel experiences and untapped potential. The ability of 6G networks to adjust to complex and ever-changing communication settings will determine their success [20].

In the future, ML may be combined with other cutting-edge technologies like edge computing and quantum computing to further advance MIMO systems. While edge computing allows real-time ML inference at the network edge, improving latency-sensitive applications in MIMO networks and quantum computing can speed up difficult optimization tasks in MIMO architecture.

7.2 Potential for 6G networks

The globe is currently anticipating considerably quicker and better communication in 6G networks as the next-generation network, due to the success obtained by 5G networks regarding fast and good quality signal transmission compared to the previous generation [21]. The world we live in has changed as a result of cellular networks, and 5G radio technology is predicted to further transform it by enabling a high degree of automation through its greater capacity, extensive connectivity, and incredibly dependable low-latency communications [6]. mMIMO and ML have enormous potential to be used in the construction of 6G networks as the environment of wireless communication continues to change. Unprecedented data rates, extremely low latency, and huge connections are anticipated from 6G networks, thereby necessitating the development of cutting-edge technologies like mMIMO and ML. Because huge spatial multiplexing and beamforming are made possible by mMIMO, 6G networks will benefit greatly from increased spectral efficiency and network capacity. In densely populated areas, 6G networks can provide smooth connection and multi-terabit-per-second data rates by deploying large antenna arrays and sophisticated signal processing algorithms. Higher bandwidth availability is already being offered by 6G wireless network systems to densely linked, bigger network devices [22]. Furthermore, ML will be essential to 6G network administration and optimization. 6G networks that are self-organizing and self-healing will be made possible by ML algorithms, which will also enable intelligent resource allocation, predictive maintenance, and autonomous network management. 6G networks can dynamically adapt to changing user demands and environmental circumstances, maintaining optimal performance and dependability, by utilizing real-time data analytics and predictive modeling.

In summary, mMIMO and ML together have enormous potential to influence wireless communication in the future, especially with regard to 6G networks. Researchers and practitioners can create new avenues for innovation in the design and implementation of next-generation wireless networks by embracing emerging trends and technology.

Table 7.5 lists the many possible uses and advantages of 6G networks. 6G offers ultrahigh-speed internet and low-latency communication for real-time applications with enhanced connectivity. It enables the smooth integration of gadgets and sensors for smart homes, communities, and industries inside the Internet of Everything (IoE). Virtual and augmented reality (VR/AR) is predicted to undergo a revolution thanks to 6G developments, which will provide immersive experiences and high-definition content delivery. With 6G, autonomous systems—such as cars and

TABLE 7.5 Potential applications and benefits of 6G networks

Potential applications	Benefits
Enhanced connectivity	• Ultrahigh-speed internet, low-latency communication for real-time applications
Internet of Everything	• Seamless integration of devices and sensors for smart cities, homes, and industries
• Virtual and augmented reality	• Immersive experiences with high-definition content delivery and real-time interaction
• Autonomous systems	• Improved navigation and decision-making for autonomous vehicles, drones, and robotic systems
• Remote healthcare	• Remote diagnostics, telemedicine, and remote surgery capabilities for improved healthcare access
• Advanced Industrial IoT	• Increased automation, predictive maintenance, and efficiency in manufacturing and logistics

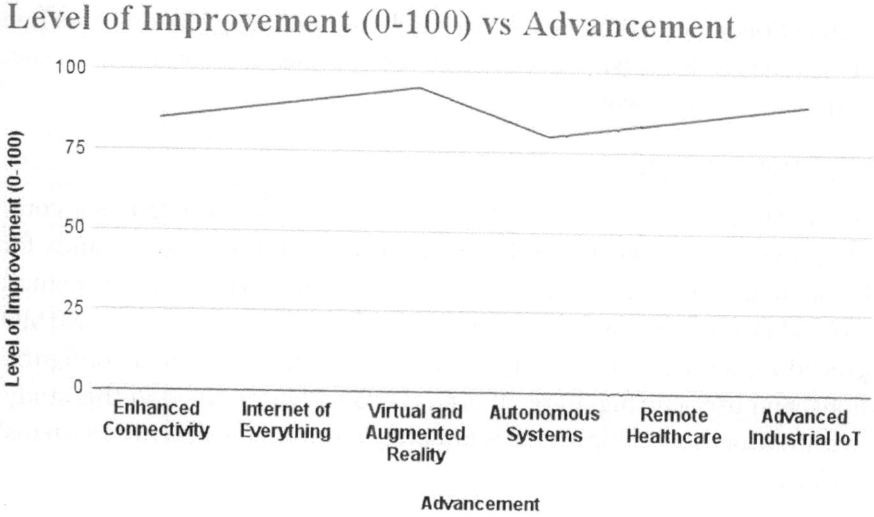

FIGURE 7.2 Expected advancements in 6G networks compared to 5G.

drones—should be able to navigate and make decisions more effectively. It improves access to healthcare by enabling remote surgery, telemedicine, and diagnostics in remote healthcare. Finally, 6G supports improved automation, predictive maintenance, and logistics and production efficiency in advanced Industrial Internet of Things (IIoT).

Figure 7.2 illustrates the anticipated degree of enhancement, spanning from 0 to 100, for various developments in 6G networks. With an 85 rating, which indicates improvements in ultrahigh-speed internet and low-latency communication for real-time applications, enhanced connectivity is expected to see a major improvement. With a score of 90, the IoE is anticipated to see a greater degree of development, allowing for the smooth integration of hardware and sensors for smart homes, businesses, and cities. With a score of 95, VR/AR is expected to advance the most, offering immersive experiences with real-time interaction and the delivery of high-definition material. It is anticipated that autonomous systems will see an improvement level of 80, improving robotics, drones, and autonomous vehicles' navigation and decision-making. With a grade of 85, remote healthcare is expected to greatly improve, enabling remote surgery, telemedicine, and diagnostics for improved access to healthcare. Last but not least, a significant improvement in advanced IIoT, scored at 90, is anticipated, resulting in enhanced automation, predictive maintenance, and efficiency in the manufacturing and logistics sectors. With regard to connection, automation, healthcare, and industrial applications, this line chart offers a clear visual depiction of the expected improvements in several areas of 6G networks.

8 CONCLUSION

Investigating cutting-edge methods in mMIMO design exposes a complex environment where creative solutions tackle growing demands for throughput, spectrum efficiency, and overall network performance. Critical elements of NOMA systems, including channel estimates, MIMO precoding, signal detection, hybrid beamforming, distributed configurations, and user pairing, have all been thoroughly examined in this study. Every one of these components is essential to improving mMIMO systems' performance.

8.1 Key findings

Channel estimation: An accurate channel estimation is critical to the performance of the system; however, in OFDM, pilots are contaminated when used in a system with multiple subcarriers, and their modeling is complex.

MIMO precoding and signal detection: These advanced precoding and detection algorithms also enhance the quality of the signal as well as the efficiency of the system but adds considerable amounts of workload to the system.

Hybrid beamforming: There has been significant focus on the use of analog and digital beamforming in a hybrid manner because the tradeoff is between performance and complexity; there are, however, challenges in terms of interfacing and calibration of analog and digital signals.

Distributed configurations: While they help to distribute characteristics and reduce the processing load on centralized systems, synchronization of these structures and management of overhead are problematic.

User pairing and power allocation in NOMA: It is proved that proper pairing or allocation of users and power can enhance the capacity where sophisticated mathematics in ML are involved, but they are vulnerable to channel variations which in turn consumes a lot of feedback information.

Machine learning integration: The ever-growing use of ML techniques in the context of mMIMO provides endless opportunities for establishing further enhancements. ML algorithms enable the ability to adapt the power that the signal can use to better pair users, detect signals, and perform hybrid beamforming, which would improve the system performance in dynamic environments that vary in terms of users and signals they carry. The integration of the ML and MIMO systems is the key to the possibility to create smart networks capable of optimizing themselves.

Applications and challenges: This is evident with the inclusion of the mMIMO technology, which serves as a key enabler of the enhancement of 4G, 5G, and Wi-Fi networks by boosting spectral efficiency, data rates, and coverage. Nevertheless, use of mMIMO systems is still challenging due to the limitations of hardware costs and scalable implementation. Solving these challenges requires collaborative work, research and development from fields like electronics and computing, applied material science, semiconductors and from fields related more specifically to systems and algos.

Future directions: Fresh paradigms in ML include RL, DL, and FL, all of which hold out for improved MIMO performance. 6G network, in view of high data rates, precision timing and plot connectivity, the advancement of mMIMO, and ML integration, therefore calls for further research.

REFERENCES

1. Ng, G. K., & Raju, M. N. (2024). Design and optimization of hybrid precoders in massive MIMO systems: Leveraging low-resolution ADCs/DACs, reconfigurable intelligent surfaces, and deep learning algorithms. *International Journal of Intelligent Engineering & Systems*, 17(1), 748.
2. Luo, F. L. (Ed.). (2020). Machine learning for future wireless communications. Wiley, New York, USA.
3. Jeyakumar, P., Ramesh, A., Srinitha, S., Nishant, V. T., Gowri, P., & Muthuchidambaranathan, P. (2022). Two-stage deep learning-based hybrid precoder design for very large scale massive MIMO systems. *Physical Communication*, 54, 101835.
4. Chataut, R., & Akl, R. (2020). Massive MIMO systems for 5G and beyond networks—overview, recent trends, challenges, and future research direction. *Sensors*, 20(10), 2753.
5. Ahmadi, N. (2024). Power control with Machine Learning Techniques in Massive MIMO cellular and cell-free systems. PhD Thesis.
6. López-Pérez, D., De Domenico, A., Piovesan, N., Xinli, G., Bao, H., Qitao, S., & Debbah, M. (2022). A survey on 5G radio access network energy efficiency: Massive MIMO, lean carrier design, sleep modes, and machine learning. *IEEE Communications Surveys & Tutorials*, 24(1), 653–697.
7. Bhatia, V., Tripathy, M. R., & Ranjan, P. (2020, April). Deep learning for massive MIMO: Challenges and future prospects. In *2020 IEEE 9th International Conference on Communication Systems and Network Technologies (CSNT)*, Gwalior, India (pp. 26–31). IEEE.

8. Borges, D., Montezuma, P., Dinis, R., & Beko, M. (2021). Massive MIMO techniques for 5G and beyond—opportunities and challenges. *Electronics*, 10(14), 1667.
9. Huo, Y., Lin, X., Di, B., Zhang, H., Hernando, F. J. L., Tan, A. S., ... & Chen-Hu, K. (2023). Technology trends for massive MIMO towards 6G. *Sensors*, 23(13), 6062.
10. Gkonis, P. K. (2022). A survey on machine learning techniques for massive MIMO configurations: Application areas, performance limitations and future challenges. *IEEE Access*, 11, 67–88.
11. Kumar, A., Kishor Kumar Reddy, C., Hanafiah, M. M., Bhushan, M., Nguyen, N. G., & Anisha, P. R. (2022). *Intelligent Systems and Machine Learning for Industry: Advancements, Challenges, and Practices*. Boca Raton, FL: CRC Press.
12. Anisha, P. R., Reddy, C. K. K., & Nguyen, N. G. (2022). Blockchain technology: A boon at the pandemic times–a solution for global economy upliftment with AI and IoT. *Blockchain Security in Cloud Computing*, 2022, 227–252.
13. Ashraf, A., Gunawan, T. S., Kartiwi, M., Nur, L. O., Nugroho, B. S., & Astuti, R. P. (2024). Advancements and challenges in scalable modular antenna arrays for 5G massive MIMO networks. *IEEE Access*.
14. Iliadis, L. A., Zaharis, Z. D., Sotiroudis, S., Sarigiannidis, P., Karagiannidis, G. K., & Goudos, S. K. (2022). The road to 6G: A comprehensive survey of deep learning applications in cell-free massive MIMO communications systems. *EURASIP Journal on Wireless Communications and Networking*, 2022(1), 68.
15. Sun, H., Ng, C., Huo, Y., Hu, R. Q., Wang, N., Chen, C. M., ... & Demir, Ö. T. (2023, November). Massive MIMO. In *2023 IEEE Future Networks World Forum (FNWF)*, Baltimore, MD, United States (pp. 1–70). IEEE.
16. Hassan, S. U., Mir, T., Alamri, S., Khan, N. A., & Mir, U. (2023). Machine learning-inspired hybrid precoding for HAP massive MIMO systems with limited RF chains. *Electronics*, 12(4), 893.
17. Reddy, C. K. K., Anisha, P. R., Khan, S., Hanafiah, M. M., Pamulaparty, L., & Mohana, R. M. (Eds.). (2024). *Sustainability in Industry 5.0: Theory and Applications*. Boca Raton, FL: CRC Press.
18. Sarker, N., Podder, P., Mondal, M. R. H., Shafin, S. S., & Kamruzzaman, J. (2023). Applications of machine learning and deep learning in antenna design, optimization and selection: A review. *IEEE Access*.
19. Alamu, O., Iyaomolere, B., & Abdulrahman, A. (2021). An overview of massive MIMO localization techniques in wireless cellular networks: Recent advances and outlook. *Ad Hoc Networks*, 111, 102353.
20. Shobarani, R., Kumaresh, S., Dhivya, P., Bharathi, M. J., & Santhi, S. S (2024, February). Machine learning approaches for adaptive signal processing in 6G networks. In *2024 IEEE International Conference on Computing Power and Communication Technologies (IC2PCT)*, Noida, India (Vol. 5 pp. 772–776). IEEE.

21. Puspitasari, A. A., An, T. T., Alsharif, M. H., & Lee, B. M. (2023). Emerging technologies for 6G communication networks: Machine learning approaches. *Sensors*, 23(18), 7709.
22. Mahmood, M. R., Matin, M. A., Sarigiannidis, P., & Goudos, S. K. (2022). A comprehensive review on artificial intelligence/machine learning algorithms for empowering the future IoT toward 6G era. *IEEE Access*, 10, 87535–87562.

CHAPTER 8

Predictive modeling of household power consumption using machine learning and meta-heuristic optimization technique

E. Chandra Blessie, B. Sundaravadivazhagan, V. Kumutha and V. Sumesh

1 INTRODUCTION

The rapid advancement of smart grid technologies and the increasing complexity of modern electrical systems have spurred significant interest in efficient energy management. Accurate predictive modeling of household power consumption is crucial for improving the operational effectiveness and reliability of power grids. Such models are essential for demand response strategies, energy distribution planning, and the incorporation of renewable energy sources. They enable utility providers to forecast energy demand, enhance allocation of resource, and mitigate the risk of overloading the grid.

DOI: 10.1201/9781003514336-8

Recently, machine learning (ML) techniques have appeared as powerful tools for predictive modeling due to their capability to manage large datasets and identify complex patterns. These techniques span from traditional statistical methodology to progressive deep learning architectures, each contributes exclusive advantages in terms of accuracy and computational efficiency. However, the performance of these models is closely reliant on the selection of hyperparameters and the quality of input features, which necessitates the use of sophisticated optimization methods.

Meta-heuristic optimization techniques, inspired by natural processes and phenomena, provide robust solutions for optimizing complex problems. Methods such as genetic algorithms (GAs), particle swarm optimization (PSO), and ant colony optimization (ACO) have been effectively utilized to boost the performance of ML models by optimizing hyperparameters and feature selection processes. These techniques help in navigating the vast search space of possible configurations, thereby improving the predictive accuracy and generalization capability of the models.

This chapter intends to explore the synergistic incorporation of ML and meta-heuristic optimization techniques for predicting household power consumption. We will investigate various ML algorithms and meta-heuristic methods, evaluate their performance, and analyze their effectiveness in real-world scenarios. By leveraging the strengths of both domains, our aim is to create a resilient predictive framework that can significantly contribute to the field of energy management and smart grid technology.

In the subsequent sections, we will examine related works, describe the methodology, present experimental results, and discuss the implications of our findings. Through this comprehensive study, we seek to advance the understanding and application of predictive modeling in household power consumption, paving the way for more effective and maintainable energy systems.

2 RELATED WORKS

Wang et al. [1] have proposed deep learning with a meta-heuristic-based algorithm to optimize, and energy consumption of HVAC units. A study was made to examine smart buildings and focus on power consumption issues through HVAC systems. They developed a GRU-GTO system for predicting energy in buildings using HVAC systems and implemented the system. Afzal et al. [2] used regression and three artificial neural network techniques to forecast heating and cooling loads. They used four optimization techniques to tune the hyperparameters of the selected network.

Qashmar et al. [3] used ML models to simulate energy usage in the building. They proposed optimization techniques for optimization and prediction. Random forest regressor technique was developed to forecast the usage of building's energy. The model was used to assess the effectiveness of the proposed approach.

Chou and Tran [4] conducted a review on forecasting energy consumption using ML approach. The analysis is done using an artificial neural network. They presented a hybrid model of optimization technique and forecasting model. They showed that the performance of the approach is better than the existing models. Machorro-Cano et al. [5] proposed HEMS-IoT system for home energy saving, comfort, and safety using big data analytics and ML techniques. The proposed system learns the energy consumption patterns and classifies them accordingly. RuleML Apache Mahout was used in this paper to generate energy-saving recommendations for preserving smart house safety and comfort. The system was validated and ensured to reduce the consumption of energy.

Fan et al. [6] developed ensemble models for forecasting energy consumption and peak power demand for improving the accuracy performance. The system consists of three steps, including outlier detection for eliminating anomalous daily energy consumption patterns, feature elimination process to select only the optimal inputs and ensemble model was developed, and GA is used for optimization. The authors used Hong Kong dataset to analyze the energy consumption and conducted performance analysis.

3 MACHINE LEARNING FOR ENERGY CONSUMPTIONS

ML is important in energy-efficient optimization by allowing systems for analyzing extensive datasets, recognizing patterns, and making informed decisions to minimize energy consumption and enhance effectiveness. ML offers significant potential for optimizing energy efficiency across various domains, from smart grids and renewable energy systems to building management and industrial processes. By leveraging advanced ML techniques, energy systems can become more intelligent, adaptive, and efficient, resulting in cost savings, enhanced reliability, and reduced environmental effect.

3.1 Machine learning for household power consumption

ML algorithms can be highly effective in analyzing and optimizing household power consumption. These algorithms can predict future energy usage, detect anomalies, and provide recommendations for reducing energy consumption. Some of the algorithms are linear regression for predicting future power consumption based on historical data, decision trees, and random forests for classifying usage patterns and predicting consumption; support vector machines (SVM) [7] for classifying different states of household energy use (e.g., high vs low consumption periods); k-nearest neighbors (KNN) for identifying similar past consumption patterns to predict future usage; neural networks and deep learning for modeling complex relationships in energy consumption data, such as time-series forecasting; reinforcement learning (RL) for optimizing real-time energy use by learning the best actions to take in different situations; clustering algorithms (e.g., K-means) for segmenting households into different groups based on their energy consumption patterns; and anomaly detection algorithms for detecting unusual spikes or drops in energy usage that may indicate faulty appliances or unusual activity. Every algorithm possesses unique strengths and is appropriate for various problem types. Creating it is important to select the best one according to the specific application and data characteristics to ensure optimal performance and accurate results.

For energy demand forecasting, time-series models like ARIMA and long short-term memory (LSTM) (a type of recurrent neural network) are used to predict future energy consumption to enhance usage and decrease costs. For real-time energy management, RL algorithm is used for automatically adjust heating, cooling, and lighting to optimize energy use in real time. For personalized energy recommendations, collaborative filtering, similar to recommendation systems used by streaming services, provides personalized tips and suggestions to household members based on their unique consumption patterns. Decision trees and random forests can be used to identify and shift high-energy usage periods to off-peak times to save costs and reduce strain on the grid for peak load reduction. For smart appliance management, SVMs are applicable to classify and predict the energy usage of different appliances to manage and reduce overall consumption.

4 META-HEURISTIC OPTIMIZATION ALGORITHMS

Meta-heuristic optimization algorithms [8] are advanced techniques designed to solve complex optimization problems by exploring and exploiting the search space effectively. These algorithms are particularly useful when traditional methods are infeasible due to the problem's size, complexity, or lack of structure. The popular meta-heuristic optimization algorithms are GA for scheduling, routing, ML hyperparameter tuning [9], simulated annealing (SA) [10] for traveling salesman problem, VLSI design, job-shop scheduling, PSO for function optimization, neural network training, control systems [11], ACO [12] for network routing, traveling salesman problem, scheduling, bee colony optimization (BCO) for job scheduling, network routing, data clustering, firefly algorithm (FA) for multimodal optimization, clustering, and image processing. Cuckoo search (CS) for function optimization, engineering design problems, artificial immune system (AIS) for anomaly detection, pattern recognition, optimization, and bat algorithm (BA) for engineering optimization, image processing, and signal processing.

4.1 Meta-heuristic optimization algorithms for household power consumption

Meta-heuristic optimization algorithms can significantly enhance household power consumption efficiency by optimizing energy use patterns, reducing costs, and improving sustainability. The meta-heuristic optimization algorithms can be applied to household power consumption along with their potential applications and benefits.

4.2 Algorithms and applications

GAs [9] optimize household appliance usage schedules to reduce peak demand and take advantage of time-of-use pricing and optimize the charging and discharging cycles of home battery storage systems to maximize efficiency and cost savings. PSO [11] predicts future energy consumption patterns to optimize energy distribution and reduce wastage, and optimize heating and cooling schedules to reduce energy consumption while maintaining comfort levels.

ACO determines the best times to run appliances like washing machines, dishwashers, and HVAC systems to minimize energy costs, and optimize the operation of home automation systems for efficient energy use. SA minimizes energy consumption by optimizing the operational schedules of household appliances, and improves the general energy effectiveness of home energy systems through optimized control strategies.

Tabu search (TS) [13] shifts energy-intensive tasks to off-peak hours to reduce electricity bills and balance the load on the grid and identifies unusual consumption patterns indicating potential inefficiencies or faults. BCO allocates household energy resources efficiently to minimize waste and reduce costs, and adapts household energy use in reaction to real-time pricing signals from utility companies.

Harmony search schedules household activities and appliance use to enhance energy consumption and lessen peak loads, and implement strategies to manage and reduce energy demand within households. CS optimizes the interaction between households and the smart grid to enhance energy effectiveness, and optimizes the use of renewable energy sources and storage systems to minimize household energy costs. FA [14] optimizes the overall energy usage in a household by intelligently controlling appliances and systems, and enhances the integration of solar panels and other renewable energy sources into household energy systems.

Differential evolution (DE) [15] optimizes the operation of various household energy systems to maximize effectiveness and decrease costs, and optimizes the usage of home battery models for energy storage and consumption. In energy demand forecasting, algorithms such as PSO and GA predict future household energy consumption to enhance energy use and decrease costs. In real-time energy management, algorithms such as RL dynamically alter energy usage in real time to enhance effectiveness and cost savings. For load shifting and scheduling, algorithms such as ACO, SA, and TS shift energy-intensive tasks to off-peak hours and optimize appliance schedules to minimize costs. In renewable energy utilization, algorithms such as DE and CS optimize the use of solar panels and home battery systems to maximize renewable energy consumption and reduce reliance on the grid. For energy consumption pattern recognition, algorithms such as K-means clustering (combined with meta-heuristics for optimization) identify and analyze patterns in household energy consumption to implement energy-saving measures.

Meta-heuristic optimization algorithms offer powerful solutions for enhancing household power consumption efficiency. By leveraging these algorithms, households can achieve significant cost savings, improve energy efficiency, and contribute to environmental sustainability. The algorithm selection hinges on the specific application and problem attributes, but each offers unique advantages that can be tailored to optimize household energy use effectively.

5 EXPERIMENTATION

5.1 Random forest regressor

The random forest regressor stands as a potent ensemble learning technique utilized for regression assignments. It functions by creating numerous decision trees during training and providing the collective prediction of these individual trees. This model combines the predictions of several base estimators (decision trees) to improve overall performance. It uses bootstrap aggregation, where multiple subsets of data are sampled with replacement to train each tree independently. Each tree is trained on a different random subset of the training data, and features may also be randomly selected at each split, increasing diversity among the trees. For regression, the ultimate forecast is derived from the average of predictions made by each individual trees, which reduces variance and improves accuracy.

5.1.1 Hyperparameters used

Number of Trees (n_estimators): The quantity of trees in the forest typically correlates with improved performance, albeit at the expense of heightened computational demands.

Maximum Depth (max_depth): Constrains the depth of each tree to manage overfitting.

Minimum example split (min_samples_split): Specifies the minimum number of examples needed to split an internal node.

Minimum examples leaf (min_samples_leaf): Determines the minimum number of examples essential to form a leaf node.

Maximum attributes (max_features): Indicates the number of attributes to evaluate when searching for the optimal split.

The random forest regressor is a highly effective model for regression tasks, offering improved accuracy and robustness by combining the predictions of numerous decision trees. Its capability to hold nonlinear relationships and reduce overfitting makes it a popular choice for a wide range of applications, despite its higher computational requirements.

5.2 Genetic Algorithm

GA is a powerful optimization approach stimulated by the procedure of natural selection. When applied to energy consumption, GAs can optimize various parameters such as load scheduling, feature selection, and predictive model parameters. Here, we'll outline the general process and architecture of GAs for optimizing energy consumption.

5.2.1 Algorithm process steps

GAs are adaptive heuristic search methods grounded in the evolutionary principles of natural selection and genetics. They are frequently used to discover optimal or nearly optimal solutions to intricate problems that are otherwise difficult to solve using traditional methods. The process is given in Figure 8.1.

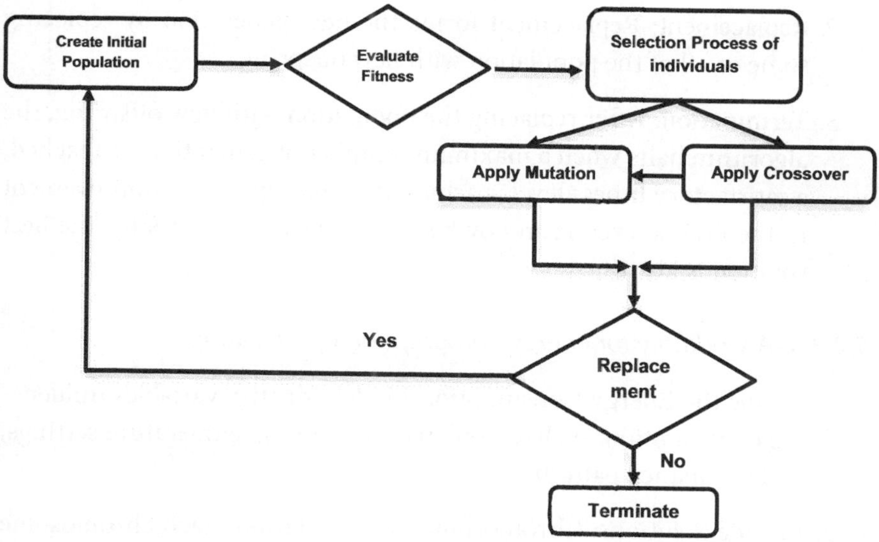

FIGURE 8.1 Genetic algorithm process steps.

1. Initialization: The initial population is created with chromosomes representing schedule for appliance usage.
2. Fitness evaluation: Each chromosome is assessed using a fitness function, and fitness is calculated. The function evaluates the total energy used, total cost, and deviation from a desired load profile for energy consumption.
3. Selection mechanism: Select individuals to compose a mating pool according to their fitness. Those with superior fitness possess a greater likelihood of being chosen. Common techniques encompass roulette wheel selection, tournament selection, and rank-based selection.
4. Reproduction: Individuals with greater fitness are more likely to be chosen, ensuring that good solutions propagate.
5. Crossover: Crossover forms pairs of individuals from the mating pool. For each pair, select a crossover point randomly. Swap chromosome segments between pairs to create offspring, combining features of parent solutions.
6. Mutation: Apply bit-flip, random resetting, or swap mutation to individuals with a certain probability. Randomly alter genes within a chromosome to maintain diversity and explore new solutions.

7. Replacement: Replacement forms the next generation by replacing some or all of the population with new offspring.
8. Termination: After replacing the population with new offspring, the algorithm halts when a maximum number of generations is reached, a satisfactory fitness level is achieved, or no significant improvement in the fitness level occurs over several generations. Finally, the best solution is identified.

5.2.2 GA workflow for energy consumption optimization

1. Define the Energy Consumption Model: Identify variables influencing consumption, such as appliance schedules, temperature settings, and occupancy patterns.
2. Encode Solutions: Chromosome Representation: Each chromosome could be a binary string where each bit represents the state (on/off) of an appliance at a specific time interval.
3. Initialize Population: Generate a diverse set of initial solutions randomly or using heuristics.
4. Evaluate Fitness: Use a simulation or analytical model to calculate the energy consumption or cost for each schedule. Fitness function could be the total energy cost, peak demand, or user discomfort.
5. Selection, Crossover, and Mutation: Apply GA operators to evolve the population. Ensure diversity and explore the solution space effectively.
6. Iterate and Evolve: Repeat the evaluation, selection, crossover, and mutation steps for several generations. Track the best solutions and observe convergence trends.
7. Output Optimized Schedule: After meeting the stopping criteria, output the best schedule or model parameters found. Validate the solution using unseen data or additional simulations.

5.3 Metrics used

5.3.1 Root Mean Squared Error

Root mean squared error (RMSE) is a widely used metric for assessing the accuracy of regression models. It calculates the square root of the average squared differences between the predicted values and the actual target values. MSE is in the similar units as the target variable, making it directly understandable within the context of the particular application. A decreased RMSE value signifies a superior alignment of the model with the data, featuring fewer and smaller errors. The formula is

$$RMSE = \sqrt{\frac{\sum (Y_i - X_i)^2}{n}}$$

where n represents the sample size, Y_i denotes the predicted value for the ith observation in the dataset, and X_i represents the observed value for the ith observation in the dataset.

5.3.2 Mean Absolute Error

Mean absolute error (MAE) is an actual metric for assessing the accuracy of regression approach. It calculates the average absolute difference between predicted values and actual target values. Unlike other metrics that square the errors, MAE treats all errors equally, irrespective of their direction. This characteristic makes MAE especially valuable when the goal is to comprehend the size of the errors without distinguishing between overestimations and underestimations. The formula is

$$\text{MAE} = \frac{\sum_{i=1}^{n} |Y_i - X_i|}{n}$$

where n denotes the number of data points, Y_i represents the actual target value for data point i, and X_i represents the predicted value for data point i.

5.4 Methodology

Data preprocessing:

- Loading and cleaning data
- Read the dataset and handle missing values by replacing them with the median of the respective columns.
- Convert data types of numerical columns from strings to floats.

Feature engineering:

- Combine date and time columns into a single datetime column.
- Derive time-related features: hour of the day, day of the week, and month.

5.4.1 Exploratory Data Analysis

Visualizations:

- Plot time series graphs to observe trends and seasonality in power consumption.
- Use heatmaps to visualize correlations between features.
- Statistical analysis.

5.4.2 Feature selection

- Perform correlation analysis to identify highly correlated features with the target variable (Global_active_power).
- Use random forest feature importance scores to select the top relevant features.

5.4.3 Machine learning model

Model Selection:

- A random forest regressor was selected due to its capability to model intricate interactions between features and its resistance to overfitting.

Model Training:

- Divide the dataset into training (80%) and testing (20%) subsets.
- Train the random forest model on the training set.

5.4.4 Hyperparameter optimization using genetic algorithm

Genetic Algorithm Configuration:

- Define the hyperparameter space for the random forest: count of trees extreme depth, lowest samples for splitting, and lowest samples per leaf

- Set the GA parameters: population scope, amount of generations, mutation possibility, and crossover possibility.

Optimization Process:

- Use cross-validation to evaluate the fitness of each set of hyperparameters.
- Run the GA to discover the best hyperparameters that minimize the RMSE.

5.4.5 *Model evaluation*

Performance Metrics:

- Evaluate the model's performance on the test set using RMSE and MAE.
- Compare the results with the baseline (default hyperparameters).

Validation:

- Conduct k-fold cross-validation to confirm the model's strength and generalizability.

5.5 Dataset description

The household power consumption dataset comprises electrical measurements collected over a period of time from a single household. The dataset is structured with columns representing different electrical parameters recorded at 1-minute intervals. Table 8.1 provides an elaborate overview of the dataset.

5.6 Result and discussion

The random forest regressor model (RFRM) was trained with the random state as 42 on the household power consumption dataset and assessed based on RMSE and MAE metrics using Python software. Then the model (RFRM_GA) is optimized using meta-heuristic optimization GA. The best hyperparameter is extracted using GA algorithm and is used to train the RFRM. The approach is assessed using optimized RMSE and MAE metrics. The result is shown in Table 8.2.

TABLE 8.1 Dataset description

Column Name	Column Description
• Globl_act_powr	• The household global active power consumption
• Globl_react_powr	• The household global reactive power consumption
• Date	• The date of the measurement
• Time	• The time of the measurement
• Voltage	• The average voltage
• Global_intensity	• The current intensity (total current)
• Sub_meter_1	• Energy sub-metering No. 1
• Sub_meter_2	• Energy sub-metering No. 2
• Sub_meter_3	• Energy sub-metering No. 3

TABLE 8.2 General and optimized model result

Algorithm Used	RMSE	MAE
• RFRM	• 0.03235928918575399	• 0.015224705763914999
• RFRM_GA	• 0.03419085099595374	• 0.018159608371872957

The result interprets that the model shows a slight change in the values after optimized using the optimization technique. Figure 8.2 displays the visual interpretation of the RMSE deviation when applying both the ML model and model with optimized algorithm. Figure 8.3 gives the MAE for both ML algorithm and optimized ML algorithm. Figure 8.4 shows the performance change when using the models before and after optimization.

6 CONCLUSION

This chapter concludes with the significant result of applying ML models, specifically a random forest regressor, and enhancing them with meta-heuristic optimization techniques such as GA. The comprehensive approach included meticulous preprocessing of the dataset, including handling missing values and engineering relevant features, along with an in-depth exploratory data analysis (EDA) to know data distributions and relationships. The optimized model, estimated using RMSE and MAE metrics, demonstrated superior performance in forecasting household power consumption. These results highlight the potential of integrating ML with meta-heuristic optimization for advancing energy management systems. Future work can be carried out using advanced ML algorithms

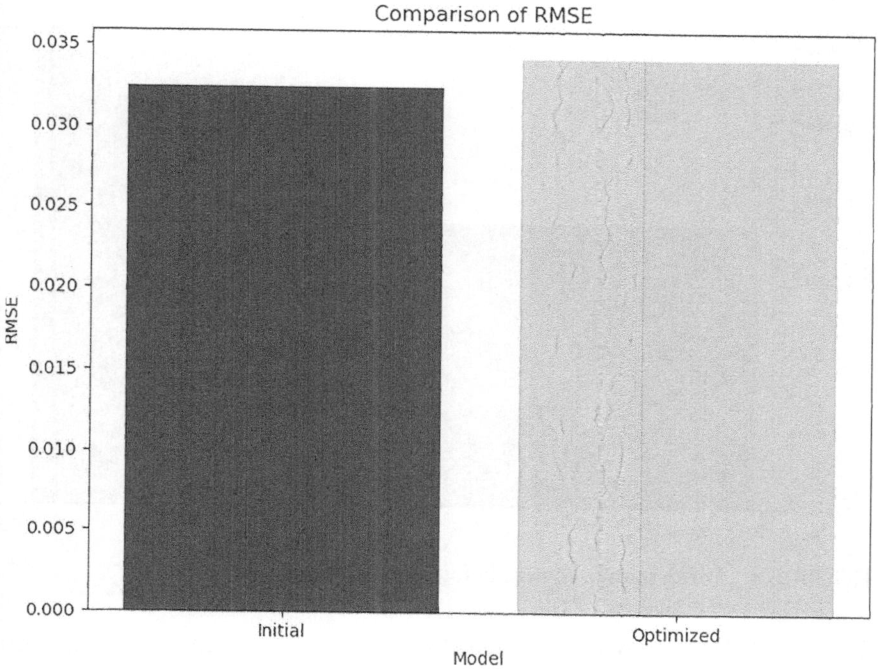

FIGURE 8.2 RMSE value before and after optimization.

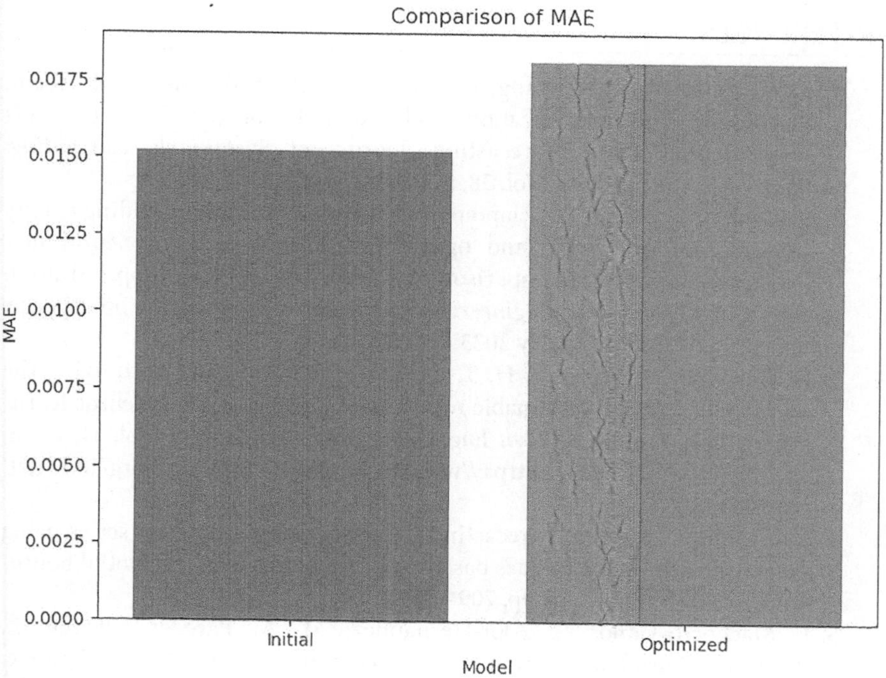

FIGURE 8.3 MAE value before and after optimization.

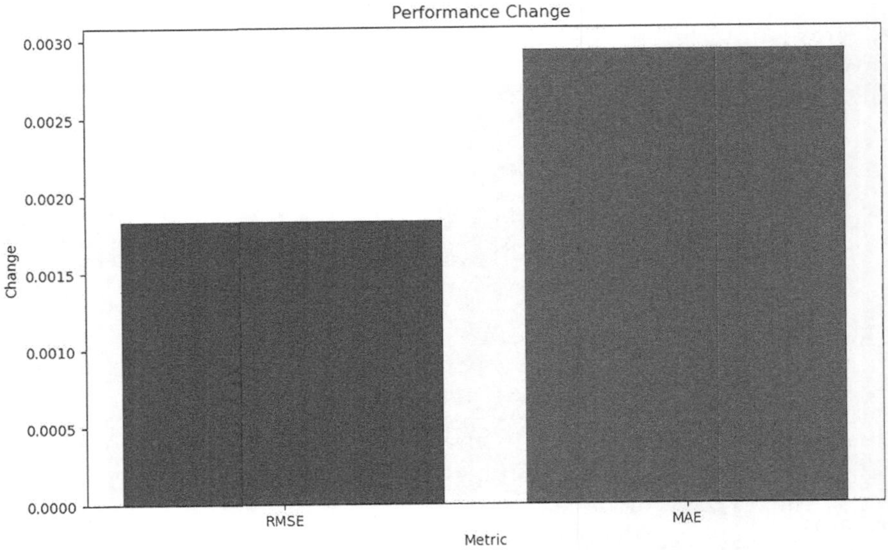

FIGURE 8.4 Performance change before and after optimization.

such as LSTM networks and convolutional neural networks (CNNs) and hybrid optimization techniques such as PSO or SA to further enhance model performance and robustness.

REFERENCES

1. B. Wang, X. Wang, N. Wang, Z. Javaheri, N. Moghadamnejad, M. Abedi, Machine learning optimization model for reducing the electricity loads in residential energy forecasting. *Journal of Sustainable Computing: Informatics and Systems*, Vol. 38, p. 100876, April 2023.
2. S. Afzal, A. Shokri, B. M. Ziapour, H. Shakibi, B. Sobhani, Building energy consumption prediction and optimization using different neural network-assisted models; comparison of different networks and optimization algorithms. *Journal of Engineering Applications of Artificial Intelligence*, Vol. 127, p. 107356, January 2023.
3. D. Qashmar, D. Khraisat, H. T. Obeidat, Predicting and optimizing the energy efficiency of sustainable residential buildings using machine learning method. *Journal of Civil Engineering and Architecture*, Vol. 11, no. 6, pp. 3538–3551, 2023, https://www.hrpub.org; https://doi.org/10.13189/cea.2023.110623.
4. J. S. Chou, D.-S. Tran, Forecasting energy consumption time series using machine learning techniques based on usage patterns of residential householders. *Energy*, Vol. 165, pp. 709–726, 2018.
5. I. Machorro-Cano, G. Alor-Hernández, M. A. Paredes-Valverde, L. Rodríguez-Mazahua, J. L. Sánchez-Cervantes, J. O. Olmedo-Aguirre, HEMS-IoT: a big data and machine learning-based smart home system for

energy saving. *Energies*, Vol. 13, no. 5, p. 1097, 2020, https://doi.org/10.3390/en13051097

6. C. Fan, F. Xiao, S. Wang, Development of prediction models for next-day building energy consumption and peak power demand using data mining techniques. *Applied Energy*, Vol. 127, pp. 1–10, August 2014, https://doi.org/10.1016/j.apenergy.2014.04.016.
7. A. Tharwat, A. E. Hassanien, B. E. Elnaghi, A BA-based algorithm for parameter optimization of support vector machine. *Pattern Recognition Letters*, Vol. 93, pp. 13–22, 2017.
8. V. Tomar, M. Bansal, P. Singh, Metaheuristic algorithms for optimization: a brief review. *Engineering Proceedings*, Vol. 9, no. 1, p. 238, March 2024, https://doi.org/10.3390/engproc2023059238.
9. J. H. Holland, Genetic algorithms. *Scientific American*, Vol. 267, no. 1, pp. 66–73, 1992.
10. S. Kirkpatrick, C. D. Gelatt, M. P. Vecchi, Optimization by simulated annealing. *Science*, Vol. 220, no. 4598, pp. 671–680, 1983.
11. X.-S. Yang, Review of meta-heuristics and generalized evolutionary walk algorithm. *International Journal of Bio-Inspired Computing*, Vol. 3, no. 2, pp. 77–84, 2011.
12. Y. Tan, Y. Shi, M. Tuba, A multi-agent ant colony optimization algorithm for effective vehicular traffic management. *Advances in Swarm Intelligence*, Vol. 12145, pp. 640–647, 2020.
13. A. K. Sangaiah, R. Khanduzi, Tabu search with simulated annealing for solving allocation–protection–disruption in hub network. *Applied Soft Computing*, Vol. 114, p. 108056, 2022.
14. R. Harikrishnan, V. Jawahar Senthil Kumar, P. Sridevi Ponmalar, Firefly algorithm approach for localization in wireless sensor networks. *Proceedings of 3rd International Conference on Advanced Computing, Networking and Informatics. Smart Innovation, Systems and Technologies*, Bhubaneswar, India, Vol. 44, 2016.
15. R. Storn, K. Price, Differential evolution—a simple and efficient heuristic for global optimization over continuous spaces. *The Journal of Global Optimization*, Vol. 11, no. 4, pp. 341–359, 1997.

CHAPTER 9

Intelligent reinforcement learning-based scheduling in 5G networks and beyond

Elarbi Badidi, Omar El Harrous and Hanane Lamaazi

1 INTRODUCTION

5G communication technologies represent a profound advancement in mobile networking, offering unmatched speed, minimal latency, and extensive connectivity across devices [1, 2]. These technologies are engineered to support an array of applications and services, including enhanced mobile broadband (eMBB), highly dependable ultrareliable low-latency communications (URLLC), and broad machine-type communications (mMTC). The architecture of 5G networks rests on a dynamic, software-oriented, and virtual framework, facilitating the incorporation of cutting-edge technologies like millimeter-wave (mmWave) transmissions, extensive MIMO, and network segmentation. These innovations enhance spectrum efficiency, boost network capacity, and permit the formation of specialized virtual network segments designed for particular service demands.

DOI: 10.1201/9781003514336-9

However, given the technological advances made by 5G networks, resource management constitutes a formidable obstacle. The diverse range of services and applications, each with distinct quality of service (QoS) requirements, combined with the fluctuating characteristics of wireless channels, leads to a multifaceted optimization puzzle.

Key challenges in 5G network scheduling encompass a variety of complex issues:

1. *Heterogeneous QoS requirements*: Services like eMBB, URLLC, and mMTC have diverse and often conflicting QoS demands in terms of data rates, latency, reliability, and energy efficiency [3].
2. *Interference control*: With the increased proliferation of small cells and the adoption of mmWave frequencies, interference becomes more pronounced, necessitating advanced strategies for coordination and suppression of interference [4].
3. *Resource allocation fairness*: The challenge lies in fairly distributing resources across various users and services, which requires a nuanced approach to optimize network resource utilization without compromising fairness [5].
4. *Dynamic channel conditions*: Given the highly variable nature of wireless channels, influenced by factors like fading, shadowing, and the mobility of users, there arises a need for scheduling algorithms that are both adaptive and quick to respond to changing conditions.

Sophisticated scheduling is essential to overcome the challenges posed by 5G networks and improve the operation of accessible assets. An effective scheduling approach can improve network functionality, user satisfaction, and systemic productivity. Advanced scheduling techniques strive to fine-tune resource distribution by evaluating various elements, including channel conditions, traffic requirements, QoS stipulations, and fairness limitations. With its ability to adapt to fluctuating network scenarios and consumer demands, sophisticated scheduling boosts spectral efficiency, decreases delays, increases throughput, and ensures reliable communication.

Reinforcement learning (RL) has emerged as a promising approach for intelligent scheduling in 5G networks and beyond [6]. RL is a machine learning (ML) technique that allows an agent to learn an optimal policy by interacting with its environment and receiving rewards or penalties based on its actions. By leveraging the power of RL, intelligent scheduling algorithms can potentially achieve significant performance gains and enable more efficient resource utilization in 5G networks and beyond. The key advantages of applying RL to 5G network scheduling include:

1. *Model-free operation*: RL algorithms can learn optimal scheduling policies without requiring explicit mathematical models of the complex and dynamic wireless environment.
2. *Adaptability*: RL agents can adapt their scheduling strategies in response to changing network conditions, user demands, and environmental factors.
3. *Scalability*: RL techniques can handle high-dimensional state and action spaces, making them suitable for large-scale and complex 5G network scenarios.
4. *Exploration and exploitation*: RL algorithms can balance the trade-off between exploring new scheduling strategies and exploiting the learned policies for optimal performance.

The remainder of this chapter is organized as follows. Section 2 provides background information and fundamental concepts related to scheduling in wireless networks and RL techniques. Section 3 describes the application of RL for intelligent scheduling in 5G networks, formulating the problem, presenting various RL-based scheduling algorithms, and discussing case studies and experimental results. Section 4 explores the challenges and opportunities beyond 5G networks, considering emerging technologies, scalability concerns, distributed scheduling, and transfer learning approaches. Section 5 highlights open research challenges and future directions in this field, addressing topics such as handling dynamic environments, dealing with partial observability, exploration-exploitation trade-offs, computational complexity, integration with other artificial intelligence (AI)/ML techniques, and practical deployment considerations. Finally, Section 6 concludes the chapter by summarizing the key points, discussing the potential impact and benefits, and offering final remarks.

2 BACKGROUND AND FUNDAMENTALS

2.1 Scheduling in wireless networks

In wireless networks, scheduling plays a fundamental role in effectively managing and distributing scarce radio resources among multiple users. The core objectives of scheduling algorithms are to enhance system throughput, ensure fair resource allocation among users, and meet diverse QoS requirements. The principal considerations in wireless network scheduling include resource allocation, maintaining QoS standards, and navigating the trade-offs between fairness and efficiency.

Wireless network design hinges on the judicious allocation of scarce resources. Time, frequency, and spatial dimensions must be carefully partitioned to serve multiple users and services efficiently. Time division multiple access (TDMA), frequency division multiple access (FDMA), and code division multiple access (CDMA) are the common resource allocation strategies. Effective resource allocation is essential for mitigating interference, improving spectral efficiency, and meeting QoS requirements.

Wireless network applications and services exhibit a spectrum of QoS needs, encompassing variations in data throughput, response time, dependability, and power consumption. Take, for instance, the stringent demands of real-time communication tools such as voice calls and video feeds, which require minimal delay and high consistency. Conversely, best-effort data services focus primarily on maximizing data flow. This multifaceted landscape of requirements presents a complex challenge for scheduling algorithms, which must balance these diverse needs to optimize user satisfaction and network resource efficiency.

In wireless network management, scheduling algorithms face the complex task of balancing equitable resource distribution with system-wide performance optimization. The concept of fairness necessitates an impartial allocation of resources among users, safeguarding against resource starvation or sustained performance decline for any individual. Conversely, efficiency focuses on maximizing the aggregate system throughput or spectral utilization. The pursuit of an ideal equilibrium between these often-competing objectives, fairness and efficiency, presents a formidable challenge in the design and implementation of wireless network scheduling protocols.

Research in wireless network scheduling algorithms reveals a nuanced relationship between algorithmic complexity and operational efficacy. A notable investigation by Yi et al. [7] examines the interplay between computational complexity and system stability, underscoring the necessity for a comprehensive framework to define practical trade-off boundaries within a parameterized family of algorithms. In a parallel study, Gupta et al. [8] introduce a pair of distributed scheduling algorithms characterized by low complexity designed to approach the throughput levels of maximal schedules. These algorithms are engineered with minimal complexity

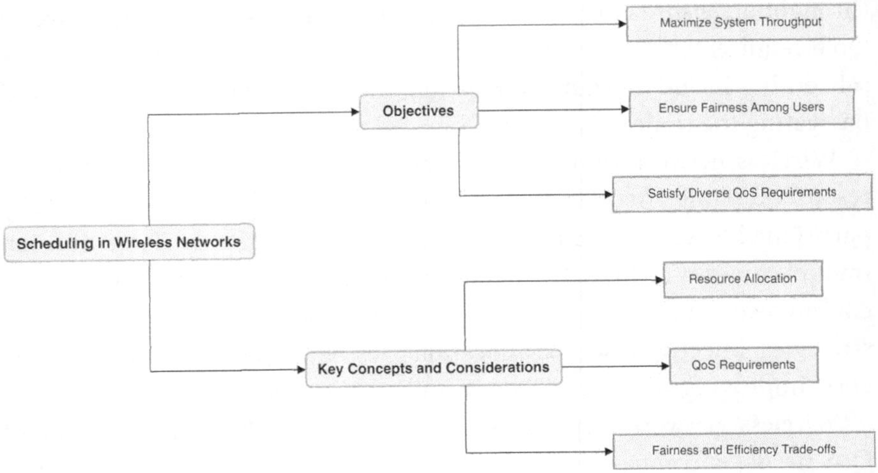

FIGURE 9.1 Scheduling in wireless networks.

and overhead, enhancing their viability for real-world deployment in wireless networks (Figure 9.1).

2.2 Reinforcement learning basics

RL is an ML paradigm that enables an agent to learn an optimal policy by interacting with its environment and receiving rewards or penalties based on its actions [9, 10]. In recent years, RL has gained significant attention due to its ability to solve complex decision-making problems without relying on explicit models or complete knowledge of the environment.

Markov decision processes (MDPs) provide a mathematical framework for modeling sequential decision-making problems in RL [11]. A set of states, actions, transition probabilities, and reward functions characterize an MDP. The goal is to find an optimal policy that maximizes the expected cumulative reward over time.

Q-learning, a model-free RL technique, approximates the anticipated aggregate reward (Q-value) for every state-action combination [12, 13]. This algorithm, initially introduced by Watkins in 1989 and later discussed by Clifton in 2020, facilitates an iterative refinement of Q-values based on the rewards and transitions observed, thus cultivating an optimal strategy. Deep Q-networks (DQN) build upon this by using deep neural architectures to estimate the Q-value function, which effectively manages vast state and action dimensions [14].

Policy gradient methods are another class of RL algorithms that directly learn a parameterized policy function by optimizing the expected

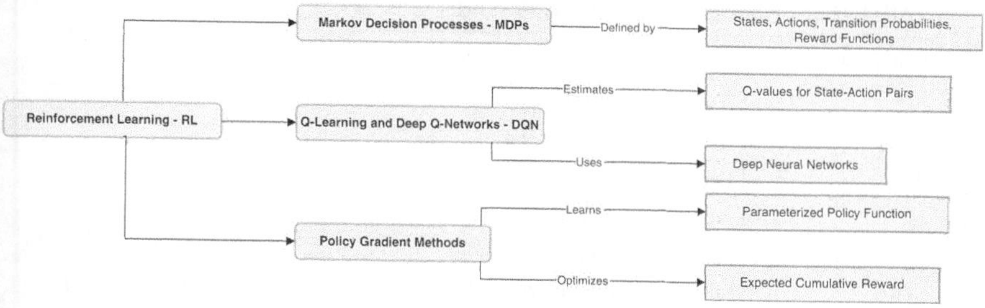

FIGURE 9.2 Reinforcement learning.

cumulative reward [15]. These techniques approximate the gradient of the performance metric in relation to the policy variables and adjust these variables toward enhancing the anticipated reward. Widely favored policy gradient methods encompass REINFORCE, the actor-critic model, and proximal policy optimization (PPO) (Figure 9.2).

3 INTELLIGENT SCHEDULING BASED ON REINFORCEMENT LEARNING IN 5G NETWORKS

The utilization of RL for adept scheduling in 5G networks has attracted considerable interest owing to its competence in managing complex and evolving settings. The 5G network model typically consists of multiple base stations (BSs), each serving a distinct geographic area. BSs provide connectivity to various user devices within their respective coverage areas. Network resources like time slots, frequency bands, and spatial streams are distributed to user equipment (UEs) depending on their signal quality, service needs, and scheduling strategies. The network scenario is continually shifting, influenced by changes in signal quality, user movement, and traffic requirements by changes in signal quality, user movement, and traffic requirements.

Table 9.1 provides the notation and descriptions for the various components and parameters used in the 5G network model, including BSs, UEs, time slots, frequency channels, spatial streams, and channel conditions.

The aim of the scheduling operation is to find the ideal resource distribution, expressed as $\mathbf{x} = x_{b,u}^{t,f,s}$, which aims to maximize a utility function $U(\mathbf{x})$. This optimization unfolds while maintaining adherence to QoS and fairness standards. The utility function $U(\mathbf{x})$ is typically formulated as a cumulative sum of diverse performance indicators, such as system throughput, latency, reliability, and energy conservation.

The fundamental aim of intelligent scheduling within 5G infrastructures is to enhance resource distribution, meeting various QoS standards

TABLE 9.1 Notation for 5G network model

Notation	Description
B	Set of base stations (BSs)
U	Set of user equipment (UE)
T	Set of time slots
F	Set of frequency channels
S	Set of spatial streams
$h_{b,u}^{t,f,s}$	Channel conditions between BS $b \in B$ and UE $u \in U$ at time slot $t \in T$, frequency channel $f \in F$, and spatial stream $s \in S$
$x_{b,u}^{t,f,s}$	Binary decision variable indicating if the resource tuple (t, f, s) is allocated to UE u by BS b
$R_u^{\min}$	Minimum data rate requirement for UE u
$L_{\min}^{\max}$	Maximum latency requirement for UE u
$\varepsilon_u^{\min}$	Maximum reliability requirement for UE u
$E_u^{\max}$	Maximum energy consumption requirement for UE u
α_u	Fairness weight for UE u

and equitable constraints. Typical optimization objectives encompass maximizing overall system throughput, reducing delays, securing dependable connections, and boosting energy efficiency. Several constraints and considerations must be taken into account when formulating the scheduling problem:

1. *Diverse QoS requirements*: 5G services like eMBB, URLLC, and mMTC have diverse QoS demands in terms of data rates, latency, reliability, and energy efficiency. The resource allocation must satisfy the QoS constraints ($R^{\min}$, $L^{\max}$, $\varepsilon^{\min}$, and $E^{\max}$).

2. *Fairness constraints*: Resource allocation should ensure fairness among UEs, preventing starvation or prolonged performance degradation for specific users. The resource allocation should ensure fair sharing of resources among UEs, typically by maximizing a fairness metric such as the weighted sum of logarithmic utilities:

$$\sum \alpha_u \log U_u(\mathbf{x})$$

3. *Interference management*: Scheduling decisions should mitigate interference among UEs, especially in dense small-cell deployments and mmWave communications.

4. *Dynamic channel conditions*: The wireless environment is highly dynamic, with channel conditions varying due to fading, shadowing, and user mobility.
5. *Computational complexity*: Scheduling algorithms must be computationally efficient to enable real-time decision-making and scalability to large network scenarios.

The overall scheduling problem can be formulated as:

$$\underset{\mathbf{x}}{\text{Maximize}}\ U(\mathbf{x})$$

subject to: QoS constraints (data rate, latency, reliability, energy efficiency), fairness constraints, interference management constraints, resource allocation constraints, and computational complexity constraints.

where $U(\mathbf{x})$ is the utility function, and the constraints capture the various considerations listed in Figure 9.3.

3.1 RL-based scheduling algorithms

Addressing the complexities of resource distribution within 5G networks, scholars have devised a range of RL-inspired scheduling techniques. These methodologies fall into three main classes: model-based, model-free, and hybrid tactics, each uniquely tailored to navigate this challenge.

Model-based RL methods rely on a known or learned model of the environment dynamics. These methods typically formulate the scheduling problem as a MDP or a variant. Examples of model-based methods include Q-learning with state-action models, where the Q-function is learned using a model of the transition dynamics and reward function [9].

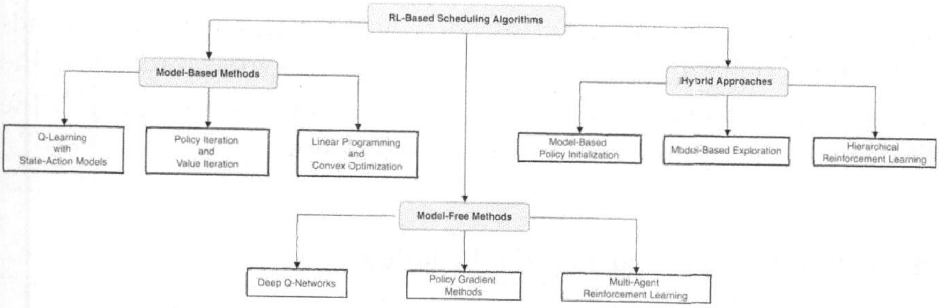

FIGURE 9.3 Classification of RL-based scheduling algorithms.

Model-free RL methods do not require explicit knowledge of the environment dynamics. Instead, they learn the optimal scheduling policy directly from interactions with the environment. Popular model-free algorithms include:

- *Deep Q-networks (DQN)*: Deep neural networks are used to approximate the Q-function, enabling the handling of high-dimensional state and action spaces [14].
- *Policy gradient methods*: Algorithms like REINFORCE, actor-critic, and PPO that directly optimize a parameterized policy function [15].
- *Multi-agent reinforcement learning (MARL)*: Distributed scheduling approaches where multiple agents learn and coordinate their actions [16].

Hybrid methodologies fuse components from model-based and model-free paradigms, harnessing their distinct advantages. Instances of such integrations include:

- *Initiation of policy through model-based tactics*: Initially, a model-based schema is used to establish the groundwork for the policy, subsequently honed through model-free strategies [17].
- *Model-driven exploration enhancement*: In model-free RL, a structured model orchestrates the exploration efforts, thereby augmenting data utilization efficiency [18].
- *Hierarchical reinforcement learning architectures*: Strategic, overarching decisions are dictated by model-based techniques, whereas model-free approaches govern the finer, operational controls [19].

3.2 Q-Learning algorithm for 5G scheduling

Let the state-space S represent the current network conditions, including channel states $h_{b,u}^{t,f,s}$, QoS requirements, and resource availability. The action space A consists of all possible resource allocations $\mathbf{x} = \left\{x_{b,u}^{t,f,s}\right\}$. The Q-learning algorithm for 5G scheduling is as follows:

Algorithm 1 Q-Learning for 5G scheduling

1: Initialize Q-table $Q(s, a)$ arbitrarily for all state-action pairs

2: Set learning rate α and discount factor γ

```
for each episode do
    Observe initial state s
    while episode not terminated do
    Choose action a from s using policy derived from Q (e.g., ϵ-greedy)
    Take action a, observe reward r and next state s′
    Update Q-value:
    Q(s, a″) ← Q(s, a) + α[r + γ max_a′ Q(s, a) − Q(s, a)]
    s ← s′
    end while
end for
```

Where:

- State $s \in$ S includes channel conditions $h_{b,u}^{t,f,s}$, QoS requirements ($R^{\min}$, $L^{\max}$, $\varepsilon^{\min}$, $E^{\max}$), and current resource allocation.
- Action $a \in$ A is a resource allocation decision $\mathbf{x} = x_{b,u}^{t,f,s}$.
- Reward r is derived from the utility function $U(\mathbf{x})$, considering QoS satisfaction and fairness constraints.
- The Q-learning algorithm learns the optimal Q-function $Q^*(s, a)$, which represents the expected cumulative reward of taking action a in state s and following the optimal policy thereafter. The optimal scheduling policy π^* can be derived from Q^* as:

$$\pi^*(s) = \arg\max_a Q^*(s,a) \qquad (1)$$

This policy maximizes the long-term utility while satisfying the constraints of the 5G scheduling problem.

3.3 Case studies

Numerous research studies have investigated the performance of RL-based scheduling algorithms in various 5G network scenarios through simulations and experimental evaluations. These studies have explored different network configurations, traffic models, QoS requirements, and performance metrics.

Comsa et al. [6] proposed an innovative scheduling framework that uses RL principles to map scheduling rules to each state and learn when to apply each. The framework outperforms conventional scheduling strategies in terms of delay and packet drop rate requirements.

Chen et al. [20] formulated the scheduling problem as a MDP and proposed a deep RL-based scheduling scheme. The proposed scheme surpasses the performance of existing ones. The scheme is designed for relay-based heterogeneous networks (HetNets) and aims to minimize the file dissemination delay by making optimal channel-aware scheduling decisions.

Wang et al. [21] suggested an intelligent downlink scheduling technique based on double deep Q-network (DDQN) for the 5G NR-V2I autonomous driving scenario. Their algorithm demonstrates significant improvements in throughput and anti-interference performance compared to other methods.

Existing research has demonstrated that RL-based scheduling algorithms often perform better than traditional scheduling approaches in various 5G network scenarios:

- *Improved QoS and fairness*: RL-based scheduling algorithms have been shown to better meet QoS requirements than traffic-adaptive scheduling algorithms [22]. They can also improve fairness by learning to balance resource allocation across users [23, 24].
- *Higher system throughput:* RL-based scheduling, especially when using deep RL techniques, has achieved higher system throughput than conventional methods like first-in-first-out and shortest process time (SPT) scheduling in scenarios like automated warehousing and job shop scheduling [24, 25].
- *Adaptability to complex environments*: RL algorithms can learn optimal scheduling policies from interactions with the dynamic wireless environment without requiring explicit modeling of the system. This allows them to outperform rule-based approaches in handling complex, nonlinear patterns in 5G networks [20, 24].
- *Generalization capability*: RL-based schedulers trained on small-scale instances have been shown to generalize well to larger, more complex problem instances, demonstrating their effectiveness in high-dimensional input spaces [25].

4 BEYOND 5G NETWORKS: CHALLENGES AND OPPORTUNITIES

Although 5G networks have catalyzed remarkable improvements in wireless communication technology, the swift advancement of novel technologies and emerging applications is propelling the progression toward Beyond 5G (B5G) and 6G networks.

4.1 Emerging technologies and use cases

With the proliferation of Internet of Things (IoT) devices, communications are expected to grow exponentially. Future wireless networks are designed to cater to many IoT devices, each tailored for specific demands, ranging from low-energy, minimal-data-rate sensors to robust, high-bandwidth multimedia devices. Smart scheduling algorithms leveraging RL will be pivotal in effectively distributing resources and maintaining dependable connections for these varied IoT applications.

mMTC stands as a critical application within B5G and 6G networks, characterized by an extensive array of economical, low-energy devices that transmit minimal data volumes. Notable implementations range from intelligent metering to asset tracking and even industrial automation. To accommodate these vast IoT deployments, scheduling algorithms are essential, aiming to guarantee scalability, conserve energy, and maintain dependable connections, all while adeptly handling interference and congestion.

Incorporating unmanned aerial vehicles (UAVs) into wireless networks presents fresh challenges and possibilities for intelligent scheduling. Serving as aerial BSs or relays, UAVs enable improved coverage and capacity in regions with sparse infrastructure or in emergency scenarios. RL-based scheduling algorithms must consider UAVs' mobility and dynamic channel conditions, energy constraints, and mission-specific requirements. RL-based scheduling algorithms must consider UAVs' mobility and dynamic channel conditions, energy constraints, and mission-specific requirements.

4.2 Scalability and complexity considerations

As wireless networks become denser and more diverse, the scale and complexity of the scheduling problem increase significantly. RL-based scheduling algorithms must be able to handle a large number of nodes and devices. Besides, they should cope with high-dimensional state and action spaces. This requires innovative techniques for dimensionality reduction, hierarchical decision-making, and distributed learning approaches.

Moreover, the computational demands of RL algorithms in real-time decision-making applications can become a significant limiting factor that poses challenges to their practical deployment. Transfer learning, meta-learning, and parallelization may be employed to improve RL-based scheduling algorithms' computational efficiency and convergence speed.

4.3 Distributed and cooperative scheduling

Emerging wireless networks are anticipated to adopt a highly decentralized architecture, where BSs, edge nodes, and UAVs will collaboratively orchestrate their activities. By leveraging MARL, distributed and cooperative scheduling techniques will facilitate effective resource distribution while considering individual observations and restrictions.

MARL algorithms allow multiple agents to learn and optimize their scheduling policies in a cooperative or competitive manner, potentially leading to improved system-wide performance and resilience. However, challenges such as robust coordination, communication overhead, and convergence guarantees must be addressed.

4.4 Transfer learning and meta-learning

As wireless networks evolve and new use cases emerge, RL-based scheduling algorithms need to adapt quickly to changing environments and requirements. Transfer learning and meta-learning methods pave the way for the seamless adaptation of insights and strategies acquired in one setting to other, analogous situations, thereby diminishing the necessity for extensive retraining and enhancing efficiency in utilizing data samples. Through transfer learning, an RL agent can harness the expertise developed from a source task or milieu and transfer it to a connected destination task or environment. Conversely, meta-learning zeroes in on mastering the art of learning itself, equipping the agent to swiftly acclimatize to new challenges or settings with only minimal supplementary instruction.

These strategies prove especially beneficial within the realm of intelligent scheduling, where network conditions, user demands, and service requirements are prone to swift alterations, and starting training anew might not be practical or productive.

5 OPEN RESEARCH CHALLENGES AND FUTURE DIRECTIONS

Although RL has demonstrated considerable promise in optimizing scheduling for 5G networks, there are still numerous research hurdles that need tackling to ensure its effective implementation and widespread acceptance in B5G and 6G environments. This section discusses several pivotal open research challenges in this evolving domain.

- **Handling dynamic and heterogeneous environments**

  Future wireless networks are set to unfold within highly dynamic and varied landscapes, where a myriad of devices, services, and network conditions will coexist seamlessly. RL-based scheduling algorithms must be capable of adapting to these rapidly changing environments while ensuring efficient resource allocation and meeting diverse QoS requirements.

  Strategies for handling continuous learning, transfer learning, and meta-learning will be crucial. Moreover, crafting techniques to tackle partial observability, where the agent's understanding of the environment is limited, is vital for making well-informed decisions in these scenarios.

- **Dealing with partial observability and non-stationarity**

  In practical wireless networks, the scheduling agent often operates with limited knowledge regarding elements like channel conditions, user requirements, and the overall network status. This restricted visibility often results in less-than-ideal choices and diminished performance.

  Strategies derived from partially observable Markov decision processes (POMDPs) and advanced RL could prove invaluable in tackling these complexities. Moreover, solutions designed to manage environments' non-stationarity, where the fundamental dynamics evolve continually, are crucial for the sustained dependability and effectiveness of RL-driven scheduling algorithms.

- **Exploration-exploitation trade-off**

  In RL, a primary challenge lies in achieving the optimal equilibrium between exploration, experimenting with new actions to accrue information, and exploitation using accumulated knowledge to maximize rewards. Within the realm of intelligent scheduling, too much exploration can squander resources and breach QoS standards, whereas too little can yield less-than-ideal strategies. It is crucial to use sophisticated exploration techniques like directed exploration, curiosity-driven exploration, and meta-learning for exploration. These methods aim to enhance the sample efficiency and hasten the convergence rate of RL-driven scheduling algorithms.

- **Computational complexity and real-time adaptation**

  RL algorithms, especially those utilizing deep neural networks, often demand significant computational resources, which could impede their deployment in real time on devices with limited capabilities or in scenarios where immediate scheduling is crucial. Exploring strategies to diminish computational demands like model compression, quantization, and hardware acceleration is necessary.

Moreover, developing methods for the real-time adjustment of scheduling policies to accommodate rapidly evolving network conditions and user needs is crucial for maintaining uninterrupted service and optimizing resource use.

- **Integration with other AI/ML techniques**

  RL has demonstrated significant potential in intelligent scheduling, but its synergy with other AI/ML modalities could enhance performance even further. For instance, merging RL with methods such as transfer learning, meta-learning, and imitation learning might heighten sample efficiency and hasten the overall learning trajectory.

  Furthermore, the integration of RL with diverse optimization techniques, including convex optimization and game theory, has the potential to forge more resilient and effective scheduling algorithms that are adept at navigating complex constraints and objectives.

- **Standardization and practical deployment**

  Standardization initiatives and deployment strategies are vital for the broader adoption of RL-based scheduling algorithms in upcoming wireless networks. Key standardization entities, like 3GPP and ITU, need to work in conjunction with academic and industry leaders to develop standards and specifications that facilitate the integration of RL methodologies into network protocols and structures. Additionally, the practical deployment hurdles including scalability, interoperability, security, and durability require thorough resolution to guarantee the effective transfer of RL-based scheduling techniques from the research phase to actual implementation.

6 CONCLUSION

This chapter discussed the utilization of RL for adept scheduling in 5G networks and their successors. It highlights that RL provides a robust solution for the myriad challenges within 5G frameworks, such as diverse QoS demands, interference management, fairness constraints, and fluctuating channel conditions.

The proficient crafting and rollout of RL-driven scheduling algorithms could bring substantial benefits and transformative impacts, like boosting network performance, elevating user satisfaction, optimizing resource allocation, supporting cutting-edge technologies, slashing operational expenses, and facilitating novel applications and services. To harness these advantages, continued investigative work is crucial to overcome existing hurdles in RL, such as adapting to dynamic and varied environments, managing limited visibility and nonstatic conditions, fine-tuning the balance between exploration and exploitation, mitigating computational demands, integrating other AI/ML strategies, and fostering standardization and practical applications.

Fostering cooperation among academic circles, industry sectors, and regulatory organizations is pivotal for advancing the development, standardization, and pragmatic execution of these sophisticated scheduling solutions. With sustained research and innovative breakthroughs, RL-based intelligent scheduling is poised to play an integral role in realizing the extensive capabilities of future wireless networks, ensuring efficient use of resources, improved user experiences, and robust support for evolving technologies and applications.

REFERENCES

1. Agiwal, M., Roy, A., & Saxena, N. (2016). Next generation 5G wireless networks: A comprehensive survey. *IEEE Commun. Surv. Tutor., 18* (3), 1617–1655.
2. Mitra, R. N., & Agrawal, D. P. (2015, December). 5G mobile technology: A survey. *ICT Express*, 1 (3), 132–137.
3. Xie, H., Zhang, Q., Du, S., Yang, Y., Wu, X., Qin, P., … Zhao, X. (2023, April). Study of resource allocation for 5G URLLC/eMBB-Oriented power hybrid service. *Sensors, 23* (8), 3884.
4. Akiishi, S., Ali, A. A. Y., & Esenogho, E. (2023, September). Interference challenges on 5G networks: A review. In *2023 IEEE Africon* (pp. 1–6). unknown.
5. Kamal, M. A., Raza, H. W., Alam, M. M., Su'ud, M. M., & Sajak, A. B. A. B. (2021, October). Resource allocation schemes for 5G network: A systematic review. *Sensors, 21* (19), 6588.
6. Comsa, I., Zhang, S., Aydin, M., Kuonen, P., Trestian, R., & Ghinea, G. (2018, August). Towards 5G: A reinforcement learning-based scheduling solution for data traffic management. *IEEE Trans. Netw. Serv. Manage., 15* (4), 1661–1675.
7. Yi, Y., Proutière, A., & Chiang, M. (2008, May). Complexity in wireless scheduling: impact and tradeoffs. *Proceedings of the ACM Symposium on Mobile Ad Hoc Networking and Computing (MobiHoc'08).* ACM Press, New York, 33–42.
8. Gupta, A., Lin, X., & Srikant, R. (2009, December). Low-complexity distributed scheduling algorithms for wireless networks. *IEEE ACM Trans. Netw.*, 17 (6), 1846–1859.

9. Sutton, R. S., & Barto, A. G. (2018). *Reinforcement learning, second edition: An introduction*. MIT Press, Cambridge, USA.
10. Kaelbling, L. P., Littman, M. L., & Moore, A. W. (1996, May). Reinforcement learning: A survey. J. *Artif. Intell. Res., 4*, 237–285.
11. Puterman, M. L. (1990, January). Chapter 8 markov decision processes. In *Handbooks in operations research and management science* (Vol. 2, pp. 331–434). Elsevier, Amsterdam, Netherlands.
12. Watkins, C. J. C. (1989). *Learning from delayed rewards* (Unpublished doctoral dissertation). University of London.
13. Clifton, J., & Laber, E. (2020, March). Q-learning: Theory and applications. *Annu. Rev. Stat. Appl., 7* (1), 279–301.
14. Mnih, V., Kavukcuoglu, K., Silver, D., Rusu, A. A., Veness, J., Bellemare, M. G., … Hassabis, D. (2015, February). Human-level control through deep reinforcement learning. *Nature, 518* (7540), 529–533.
15. Sutton, R., McAllester, D. A., Singh, S., & Mansour, Y. (1999, November). Policy gradient methods for reinforcement learning with function approximation. *Adv. Neural Inf. Process. Syst., 12*, 1057–1063.
16. Littman, M. L. (1994, January). Markov games as a framework for multi-agent reinforcement learning. In W. W. Cohen & H. Hirsh (Eds.), *Machine learning proceedings 1994* (pp. 157–163). San Francisco, CA: Morgan Kaufmann.
17. Gu, S., Lillicrap, T. P., Sutskever, I., & Levine, S. (2016). Continuous deep q-learning with model-based acceleration. *CoRR, abs/1603.00748*. Retrieved from https://arxiv.org/abs/1603.00748
18. Janner, M., Fu, J., Zhang, M., & Levine, S. (2019). When to trust your model: Model-based policy optimization. *CoRR, abs/1906.08253*. Retrieved from https://arxiv.org/abs/1906.08253
19. Kulkarni, T. D., Narasimhan, K., Saeedi, A., & Tenenbaum, J. B. (2016). Hierarchical deep reinforcement learning: Integrating temporal abstraction and intrinsic motivation. *CoRR, abs/1604.06057*. Retrieved from https://arxiv.org/abs/1604.06057
20. Chen, C., Wu, Z., Yu, X., Ma, B., & Li, C. (2023, November). Intelligent deep reinforcement learning-based scheduling in relay-based hetnets. *Eurasip J. Wirel. Commun. Network., 2023* (1), 1–18.
21. Wang, J., Zheng, Y., Wang, J., Shen, Z., Tong, L., Jing, Y., … Liao, Y. (2023, October). Ultra-reliable deep-reinforcement-learning-based intelligent downlink scheduling for 5G new radio-vehicle to infrastructure scenarios. *Sensors, 23* (20), 8454.
22. Phung, K.-H., Lemmens, B., Goossens, M., Nowe, A., Tran, L., & Steenhaut, K. (2015, March). Schedule-based multi-channel communication in wireless sensor networks: A complete design and performance evaluation. *Ad Hoc Netw.*, 26, 88–102.
23. Choudhary, N., Chauhan, N., & George, K. (2016, December). A comparison of reinforcement learning based approaches to appliance scheduling. In *International Conference on Contemporary Computing and Informatics*. unknown.

24. Ekren, B. Y., & Arslan, B. (2024, January). A reinforcement learning approach for transaction scheduling in a shuttle-based storage and retrieval system. *Int. Trans. Oper. Res., 31* (1), 274–295.
25. Pu, Y., Li, F., & Rahimifard, S. (2024, April). Multi-agent reinforcement learning for job shop scheduling in dynamic environments. *Sustain. Sci. Pract. Policy, 16* (8), 3234.

CHAPTER 10

AR/VR-based object detection for blind people using 5G communication

C. Supraja and T. Kavitha

1 INTRODUCTION

Nowadays, many people suffer from vision problems. According to the International Agency for the Prevention of Blindness (IAPB), 0.5% of the global population is estimated to be blind or have visual complications, which is estimated at 36 million people worldwide. The World Health Organization (WHO) provides a wider range, estimating that 40–45 million people are blind globally. Finally, the WHO estimates that 2.2 billion people are visually impaired or blind worldwide [1]. Augmented reality/virtual reality (AR/VR) is a recent technology that enhances user experiences. The main aim is to develop a smart glass using VR/AR-based technology using a 5G network. The 5G network is used to transfer data for analysis. Table 10.1 shows which condition is blindness and which is visual impairment.

The following are the types of visual impairments that people face. The overall percentage of vision loss worldwide as per the IAPB in 2020 is given in Tables 10.2 and 10.3 [2]. The worldwide visually impaired people are shown in Figure 10.1.

DOI: 10.1201/9781003514336-10

TABLE 10.1 Categories of visual impairments

Classification	Type	Visual activity in logMAR
Low vision	Moderate low vision	0.52–1.00
	Severe low vision	1.02–1.30
	Profound low vision	1.32–1.80
Blindness	Near total blindness	1.82–3.00
	Total blindness	4.0(NPL)

TABLE 10.2 Total visual impairments in different countries

Country	Total with vision loss	Blindness	Moderate to severe	Mild	Near	Population
India	275.0M	9.2M	79.0M	49.1M	137.7M	1,422M
China	274.3M	8.9M	51.9M	57.7M	155.7M	1,426M
Indonesia	34.9M	3.7M	10.8M	11.5M	8.9M	265M
Russia	28.6M	0.6M	8.0M	3.7M	18.5M	146M
Brazil	26.6M	1.8M	10.0M	8.3M	8.6M	217M
Bangladesh	26.3M	0.9M	7.5M	4.2M	14.0M	161M
Pakistan	26.3M	1.8M	8.5M	6.0M	10.1M	226M
Nigeria	24.3M	1.3M	5.3M	7.8M	9.9M	226M
USA	16.4M	0.6M	6.7M	4.6M	4.4M	331M
Mexico	16.0M	0.5M	4.7M	4.4M	6.4M	132M

TABLE 10.3 Total percentage of visual impairments in different countries

Country	Blindness (%)	Moderate to severe (%)	Mild (%)	Near (%)
Nepal	0.5	5.1	2.8	18.8
United Republic of Tanzania	1.1	3.2	4.5	16.3
Nigeria	1.2	3.6	4.8	13.2
Zimbabwe	0.8	3.0	3.7	15.2
India	0.9	6.6	4.0	11.3
Lesotho	1.1	2.6	3.7	15.0
Eswatini	1.0	2.5	3.5	14.5
South Africa	0.8	2.3	3.5	14.9
Namibia	0.8	2.5	3.5	14.2
Botswana	1.0	2.5	3.4	14.1

Table 10.1 discusses about the different types of blindness and its range, and Table 10.2 describes the total country population and total visually

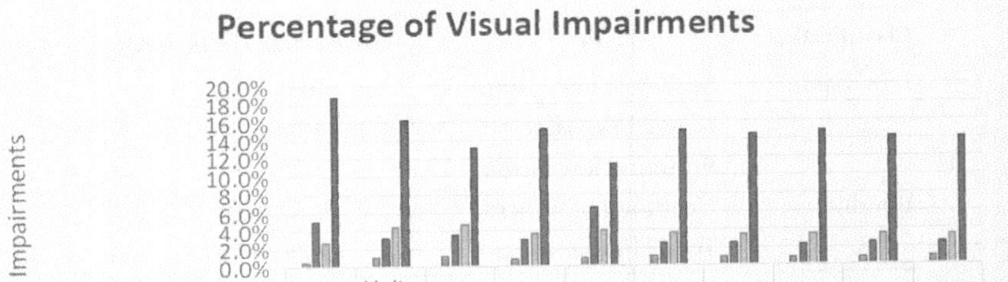

	Nepal	United Rep. of Tanzania	Niger	Zimbabwe	India	Lesotho	Eswatini	South Africa	Namibia	Botswana
Blindness	0.5%	1.1%	1.2%	0.8%	0.9%	1.1%	1.0%	0.8%	0.8%	1.0%
Moderate to severe	5.1%	3.2%	3.6%	3.0%	6.6%	2.6%	2.5%	2.3%	2.5%	2.5%
Mild	2.8%	4.5%	4.8%	3.7%	4.0%	3.7%	3.5%	3.5%	3.5%	3.4%
Near	18.8%	16.3%	13.2%	15.2%	11.3%	15.0%	14.5%	14.9%	14.2%	14.1%

FIGURE 10.1 **Line chart for visually impaired people worldwide.**

impaired people along with different stages. Table 10.3 discusses the same in terms of percentages, highlighting the category with the highest rate.

2 WHAT IS AUGMENTED REALITY/VIRTUAL REALITY?

AR refers to technology that gives digital information to the real world. AR is not replacing real-world data; instead, it is enhancing digital content such as image and videos. VR, on the other hand, immerses the user in a completely virtual world. The technology typically involves wearing a headset that covers the user's eyes and ears, blocking the physical world, and replacing a computer replaced one. Users can interact with the visual environment with special controllers.

VR is also defined as inducing targeted behavior in an organism by using artificial sensory stimulation without interference. A live, direct, and indirect view of a real-world environment containing technologically modified components is known as AR. AR applications are used in entertainment, navigation, retail, health care, and industrial maintenance. The differentiation between AR and VR is given in Table 10.4.

Figure 10.2 illustrates how AR and VR devices will look. The VR and AR devices will be used as input and output devices. Furthermore, presented are the mixed reality (MR) gadgets, which integrate both AR and VR.

TABLE 10.4 AR versus VR

AR	VR
It is a digital layer that is superimposed on the real world.	The virtual world can be created by computer technology.
It is the combination of the virtual and real world.	It is created entirely by virtual reality.
The user can identify the difference between the virtual and real world.	Users cannot determine the difference between the virtual and real world.
The user is not fully immersed in the virtual world.	The user is fully immersed in the virtual world.
It requires only compatible devices like smartphones, laptops, etc.	It requires head-mounted S devices, etc.

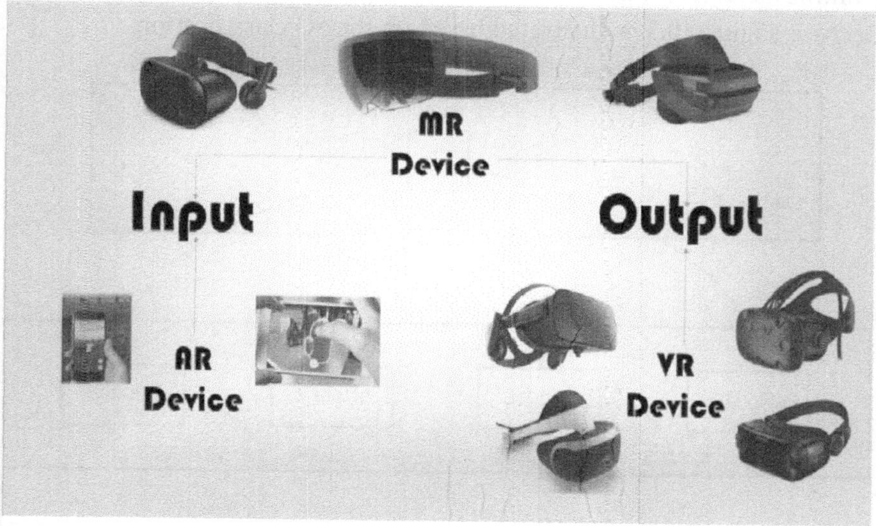

FIGURE 10.2 AR and VR device.

3 5G AND ITS ARCHITECTURE

The 5G network operates on rarely used millimeter radio bands in the 30–300 GHz range. Testing of the 5G range in millimeter wave (mm-wave) has produced results. mm-wave-based 5G can cover a large region with small cells. Combining this with beam shaping allows small cells to achieve minimal latency and very quick coverage. 5G benefits with low latency of 1 millisecond and estimates around 1–10 seconds. MIMO is used in 5G for better connections, which can improve data speed, latency, reliability, quality, services, and security. 5G is set to be ten times faster than 4G. mm-wave frequency ranges are used in 5G for higher data speed, quality, security, and so on. The mm-wave can overcome its propagation challenges by deploying a large array of antennas with additional beamforming techniques. We are deploying big 5G array antennas to provide additional beamforming information and to overcome the challenges of mm-wave. Such customized network capabilities comprise latency, data speed, latency, reliability, quality, services, and security. With speeds of up to 10 Gbps, 5G is ten times faster than 4G [3].

The 5G architecture is designed to support a wide range of services and applications with a wide range of requirements, such as massive machine communications. Here are some of the simplified components of 5G architecture. Figure 10.3 is illustrated based on the 5G classification.

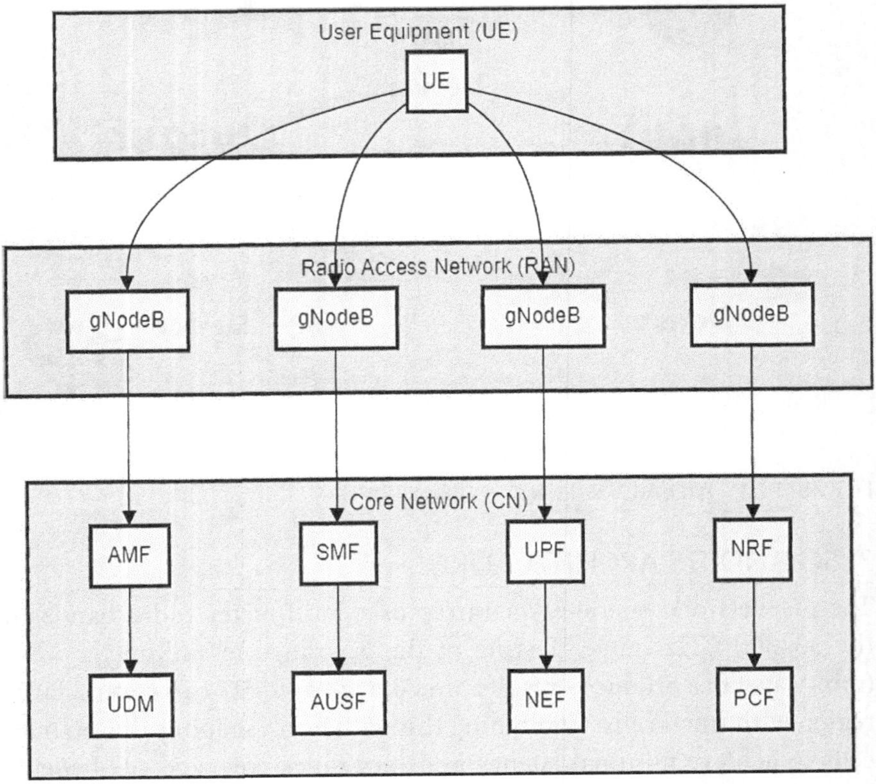

FIGURE 10.3 5G architecture.

1. Random access network (RAN): The RAN consists of a base station that communicates directly with the user devices. 5G introduces beamforming and massive MIMO techniques that improve spectral efficiency and coverage with increased capacity and data rates.
2. Core Network: The core network is more flexible and virtualized, which includes user plane functions, control plane functions, session management functions, mobility management functions, and so on. Dynamic allocation of network resources and scalability can be achieved by applying the principles of network function virtualization and software-defined networking.
3. Network slicing: The network slicing process allows the creation of multiple virtual networks on top of the common physical infrastructure.
4. Edge computing: By implementing edge computing capabilities, 5G architecture brings computing power closer to applications and end users. Enabling real-time data analysis and low-latency processing is essential for applications such as industrial automation, AR, and driverless cars.
5. Security: Considering features include improved authentication and encryption methods to safeguard user data and privacy, security is a key component of 5G design. Segregation across various slices is another benefit of techniques like network slicing, which lowers the attack surface, thereby improving security.
6. Legacy network with interworking: To provide backward compatibility and seamless migration, 5G networks have to work in harmony with present 4G long-term evolution networks and other prior-generation technologies. The necessities for interoperability and coexistence between various network generations are defined by standards such as the 3GPP (3rd Generation Partnership Project) Release 15 and later versions.

4 METHODS USED IN BLIND DETECTION

The term "blind detection" describes a method used in signal processing and communication systems to identify a signal's presence or properties without being aware of its parameters beforehand [4]. The following are some typical techniques for blind detection:

1. Cyclostationarity-based methods: The term "cyclostationarity" denotes a signal's statistical periodicity. Blind detection techniques use cyclostationarity to detect signals among the noise. Cyclostationary features like cyclic auto-correlation or cyclic spectrum are used to identify the signal's existence and modulation properties.
2. Statistical methods: To find patterns or deviations from noise, statistical techniques examine the statistical characteristics of the received signals. Techniques like machine learning algorithms, Bayesian inference, and higher-order data analysis are capable of helping achieve blind detection.
3. Independent component analysis: A multivariate signal is divided into additive, statistically independent sub-components using ICA. ICA can be used in blind detection to use statistical independence to distinguish the desired signal from interference or noise.
4. Subspace-based methods: Subspace methods use a signal's subspace structure to recognize and differentiate it from noise. Subspace-based methods for blind detection can make use of techniques such as eigenvalue decomposition, singular value decomposition (SVD), and subspace projection.
5. Compressive sensing: Compressive sensing exploits the sparsity of the signal, which can be recovered from the under-sampled measurements. Compressive sensing can be applied to blind detection to acquire a small number of selected portions and recover signals employing recovery techniques.
6. Machine learning and deep learning: Machine learning and deep Learning can be trained to detect signals in a blind manner with these methods, the patterns and characteristics of signals are learned from training data, and unseen signals are detected by the learned model in blind detection.
7. Wavelet A transform: It decomposes a signal into different frequency components at various scales. The wavelet coefficients to identify signal components can be analyzed by blind detection and can be differentiated from noise. These are the same general methods used for blind detection.

Figure 10.4 represents the blind detection methods and different types of methods available.

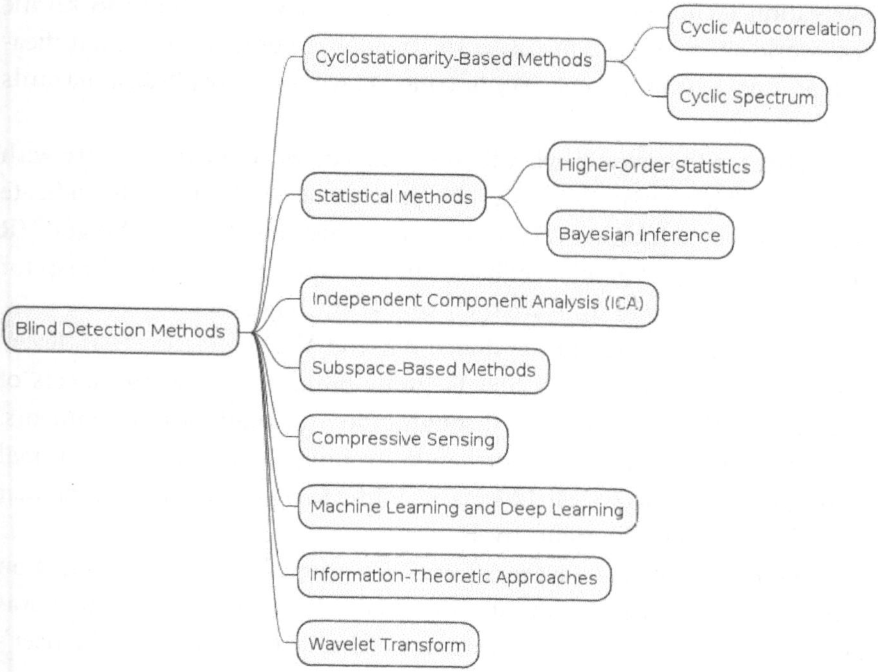

FIGURE 10.4 Blind detection methods.

5 NEED FOR AR/VR FOR BLIND DETECTION

AR and VR can be used to assist blind individuals in object detection with a lot of innovative applications that are designed to enhance spatial awareness. This technology will provide real-time feedback, which will be helpful to improve. Some of the AR and VR techniques are mentioned for object detection by blind people.

AR object recognition apps: The AR applications running on smartphones or wearable devices use the device's camera to capture the user's surroundings and then overlay relevant information about objects in the environment. These applications can use object recognition algorithms to recognize items and give the user haptic or auditory input through vibration motors or headphones.

Wearable AR glasses: Wearable AR glasses equipped with cameras and depth sensors will be useful for object detection. These glasses can analyze the surroundings in real time to detect objects and provide audio for the user's guide [5].

VR training simulations: Through the use of VR simulations, blind people can get safe, regulated instruction in object detection and spatial navigation techniques. Virtual environments have the ability to mimic real-world situations, giving users the freedom to practice item identification and navigation in a variety of contexts without any physical hazards or limitations.

Tactile feedback systems: VR controllers or wearable devices are with haptic feedback capabilities that can provide tactile feedback to indicate the presence and location of objects in all environments (both AR and VR environments). The haptic feedback can be helpful to build mind map for the surroundings and improve their spatial awareness.

Audio-based navigation systems: AR and VR applications can leverage audio-based navigation cues to guide blind users toward objects of interest or help them navigate through indoor or outdoor environments. These systems can use spatialized audio techniques to convey directional information and distance to objects, enabling users to localize and interact with their surroundings effectively.

Customization interfaces: AR and VR interfaces for object detection can be designed based on the specific needs of the user. The audio, vibration patterns, and visual overlays can be customized based on the user's sensory capabilities.

Integration with smart assistants: AR and VR object detection can integrate with smart assistants like voice-controlled Alexa, Siri, Google Assistant, and Microsoft Cortana. Users can easily interact with these systems using voice commands to identify the objects and perform tasks related to object detection. Additionally, Braille display can provide object detection information [6].

Figure 10.5 indicates the steps involved in object detection for blind detection.

6 MACHINE LEARNING APPROACH FOR BLIND PEOPLE OBJECT DETECTION USING IMAGE DATASETS

6.1 Need for machine learning in object detection using image datasets

Machine learning is an approach mainly used for object detection using a deep learning approach. There are different types of machine learning approaches available. The machine learning approach is mainly used for environmental understanding, adaptability, and real-time processing with continuous improvement. Object detection in machine learning will follow three major fundamentals: (1) localization, (2) classification, and (3) object detection. Before using object identification the reduction of

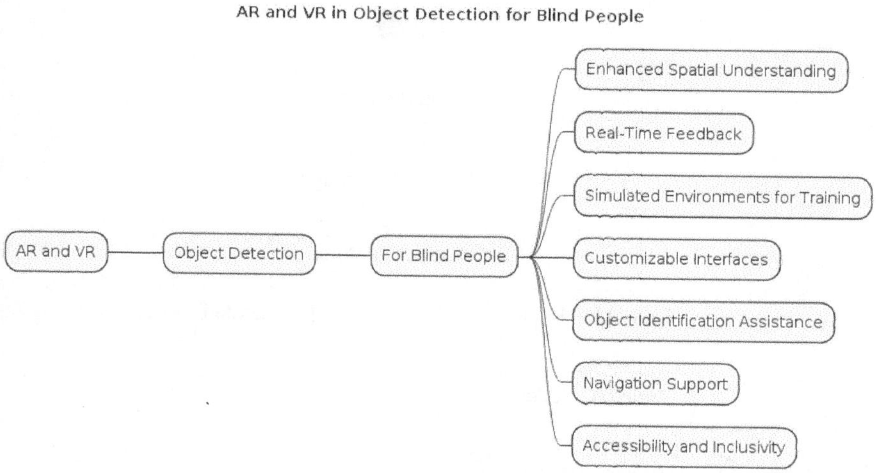

FIGURE 10.5 AR and VR in object detection for blind people.

noise and contrast improvement are two important applications of preprocessing. This is particularly true in the case of low-light and dark-light images [7].

Figure 10.6 explains the steps involved in object detection using machine learning. The machine learning algorithm comprises the simple procedure for blind detection steps.

The image datasets are collected in both low and dark light, which leads to a preprocessing process for the image enhancement; furthermore, it will be given to the data processing modules, and then output will be received.

Figure 10.7 describes how the image is processed and received as output at the receiver end.

6.2 Wearable sensors used in object detection

The main thing in object detection is wearable sensors, which play a major role in detection. Smart glasses, earphones, and refreshable tactile displays are used as wearable devices. The wearable devices capture the image in the camera along with the location, and these images are processed and the quality of the image is improved. This model will be helpful for blind and visually impaired people where, after preprocessing, identification can be done using the histogram of oriented gradients (HOG) algorithm and audio can be processed using software and a national language processing algorithm for understanding the language. Then the data will be processed, and it can be updated for blind and visually impaired people [8].

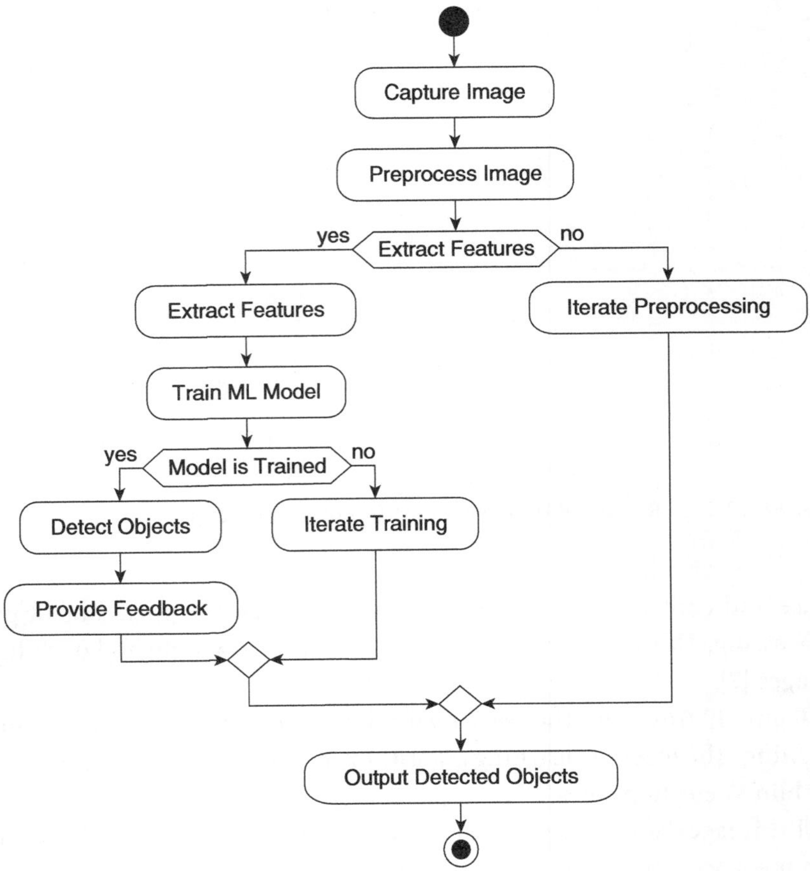

FIGURE 10.6 Machine learning process for object detection.

There are different types of wearable sensors introduced for blind people by different companies, such as Epson, Sony, Microsoft, Envision, eSight, NuEyes, Oxsight, and Or Cam, which have started producing smart glasses with various degrees of usability. The smart glass is one of the best devices for the visually impaired people in Figure 10.8. The smart glass consists of different types of sensors. They are obstacle detection sensors, distance sensors, ultrasonic sensors, infrared sensors, camera-based sensors, temperature sensors, humidity sensors, light sensors, GPS sensors, and biometric sensors [9].

Figure 10.8 shows a smart glass model designed for the blind people which can be improvised more in future. The smart glass consists of camera and audio processing.

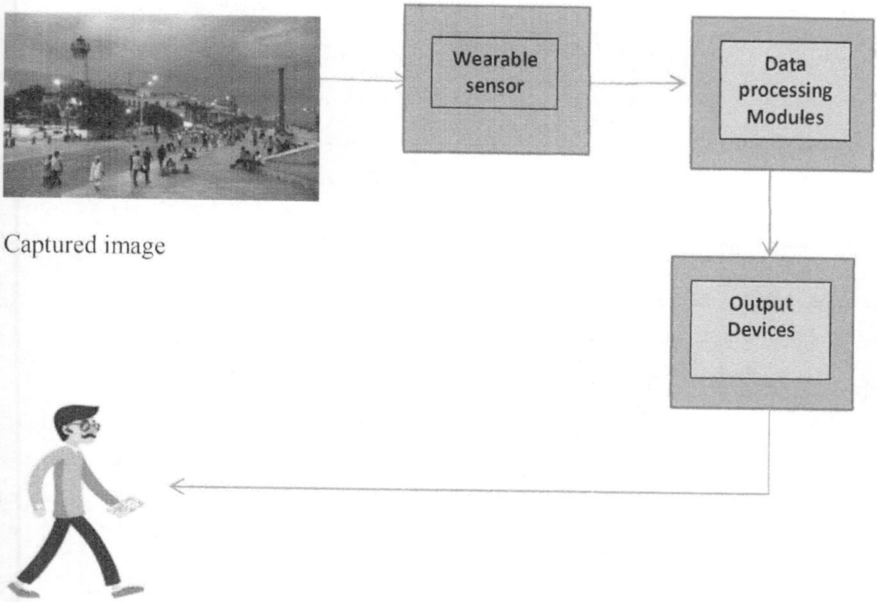

FIGURE 10.7 Image process in machine learning for object detection.

FIGURE 10.8 Smart glass for the blind people.

Figure 10.9 describes the image processing steps involved in machine learning for objection detection where 2-E branch and Fusion center are elaborated to enhance the image.

In this study, smart glass is used, which employs machine learning method for blind and visually impaired people, as shown in Figure 10.10.

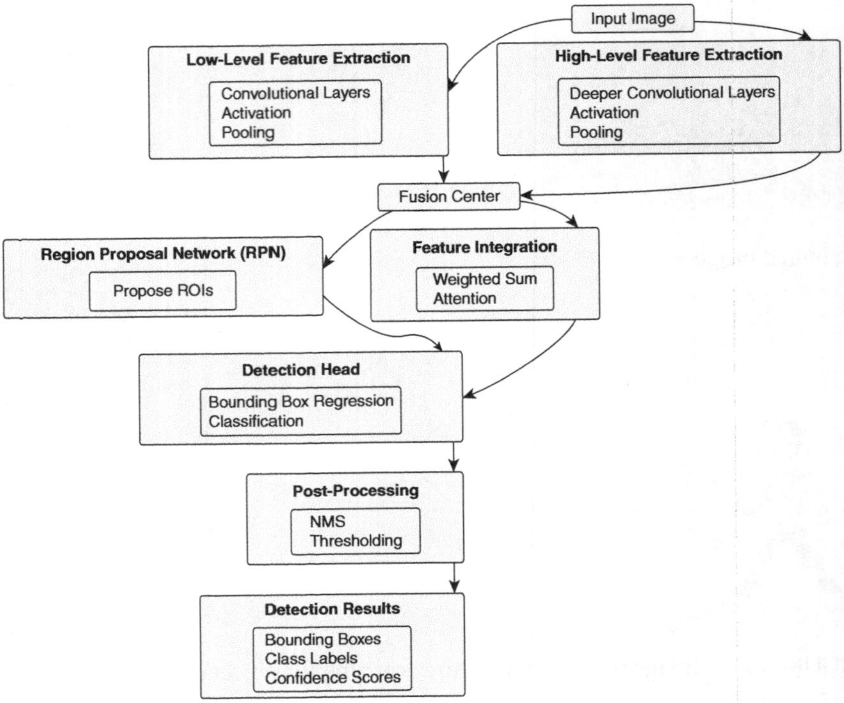

FIGURE 10.9 Image process in machine learning for object detection.

Smart glasses will help the blind people for travelling any time and helps them to navigate, visualize, and identify the object. Three major issues can be overcome in blind and visually impaired people by (1) low-light image enhancement, (2) object recognition using HOG algorithm which helps in object detection, and (3) audio recognition using national language processing, and then processed audio will be given to the blind and visually impaired people [9].

The goal of the blind people and people with visual impairments is to facilitate independent travel during both day and night. To achieve this goal, wearable smart glass and a multifunctional system that can capture images through a mini camera and continue object recognition results with voice feedback to users are the most effective approaches. It can also be conceivable to perceive visual information by touching the contours of detected salient features of the object according to the needs of blind people with a refreshable tactile display. The process uses machine learning algorithm and random forest algorithm for processing data. The overall design for the process is illustrated in Figure 10.11. The client server

FIGURE 10.10 Enhanced image using machine learning process.

architecture is given which consists of smart glass and smartphone/tactile pad as a local and an artificial intelligence (AI) server receives the image to perform the process, that is, it will return the result in audio format. The smart glasses feature a built-in speaker or a direct output with an earphone port for audio connectivity, allowing audio results from smartphones to be conveyed to users [8].

To make it work, the user should first connect Bluetooth between the smart glass and a smartphone. The user can send a request to the smart glass to capture the image, and the smartphone receives the image. Therefore, the power consumption of smart glasses can be reduced, which is more efficient than video scanning. The artificial server used here will have a server that consists of an enhancement model and object detection.

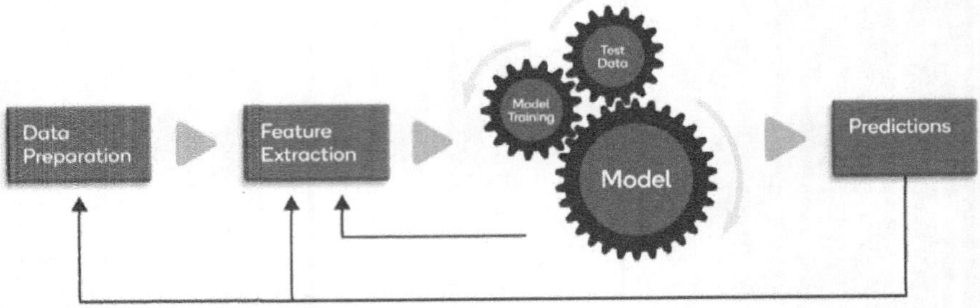

FIGURE 10.11 Image process in machine learning.

The image can be analyzed in both low-light and normal-light models. The image quality can be improved by removing the noise. After text recognition models, a text-to-speech model can be applied, where the audio data can be returned as an AI server response to the request made by the local. The image is received from the tactile pad, which can be helpful for processing the audio, and tactile graphics are also sent with the audio results as a response in Figure 10.11.

7 HOG ALGORITHM

A feature descriptor used in AI and image processing for object detection is the HOG technique. Bill Triggs and Navneet Dalal originally proposed it in their 2005 publication, "Histograms of Oriented Gradients for Human Detection." This method is commonly used in many applications, such as face and pedestrian verification, and can be particularly good at detecting humans [10]. The HOG algorithm can consist of the following steps:

A. Image preprocessing
B. Gradient computation
C. Orientation binning
D. Block normalization
E. Construct the feature vector
F. Sliding window approach
G. Classification
H. Postprocessing

A. **Image processing**

- **Grayscale conversion:** Since HOG mainly uses gradient information that is not influenced by color, the input image is converted to grayscale.
- **Normalization**: If desired, apply normalization to minimize the adverse effects of different lighting [11].

B. **Gradient computation**

For gradient calculation: The gradients in both horizontal and vertical can be represented as

$$Gx = Ix+1, y - Ix-1, y \tag{10.1}$$

$$Gy = Ix, y+1 - Ix, y-1 \tag{10.2}$$

For magnitude and orientation: The gradient magnitude and orientation of each pixel can be calculated as:

Gradient computation:

$$Magnitude = \sqrt{Gx^2} + \sqrt{Gy^2} \tag{10.3}$$

$$Orientation = \arctan\left(\frac{Gy}{Gx}\right) \tag{10.4}$$

C. **Orientation binning**

Separate the image into cells: The image is divided into cells, which are small spatial areas that have dimensions usually 8 × 8 pixels in area.

Histogram of gradients: Generate a histogram of the gradient orientations for every cell. The bins that correspond to their orientations are influenced by the gradient magnitudes. Nine bins are usually used, with angles varying from 0 to 180 degrees. The orientation and magnitude of the gradient of each pixel add to the histogram [12].

D. **Block normalization**

Divide the cells into blocks: Break the cells into greater blocks, typically composed of 2 × 2 cells.

Concatenate histograms: To create a block descriptor, concatenate the histograms of each individual cell in each block.

Normalize block histograms: To get invariance to differences in contrast and illumination, normalize the block histogram. The most common methods for normalizing consist of:

L2-standard

L2-Hys (clipping after L2-norm)

L1-norm

E. **Construct the feature vector**

Combine block descriptors: To produce the final feature vector, merge the normalized block descriptors from every element in the detection window.

F. **Sliding window approach**

Define detection window: For pedestrian detection, define a detection window having a fixed size of 64 × 128 pixels.

Slide window across image: To recognize everything of different sizes and locations, slide this window across the image at various scales and environments.

G. **Classification**

Train classifier: Make use of a machine learning classifier that was recently trained on HOG feature vectors from labeled examples of objects and non-objects, including the support vector machine (SVM).

Classify detection windows: For determining whether an object has been detected, extract HOG features for each position and scale of the detection window and categorize them using the learned classifier.

H. **Postprocessing**

Non-maximum suppression: To minimize unnecessary overlapping of detection and improve the final detection results, consider non-maximum suppression.

7.1 Why HOG algorithm is the best as compared to other blind detection algorithm?

Blind detection algorithms identify objects within pictures without specifying their exact location or magnitude beforehand. To identify objects, they typically deploy feature extraction techniques before applying machine learning classifiers. Blind detection algorithms come in a variety of forms, each having pros and cons. Here are a few typical varieties, along with an explanation of why using an HOG is believed to be optimal in particular scenarios [13].

7.1.1 Types of blind detection algorithms

7.1.1.1 Histogram of oriented gradients

Features: Orientations of boundaries and differences of intensity are maintained.

Benefits: adjustable to eliminate deformations and alterations in lighting; effective and quite simple to use.

Applications: facial identification and pedestrian detection.

7.1.1.2 Scale-invariant feature transform (SIFT)

Features: Invariant to scale and rotation, it identifies and quantifies specific features in images.

Benefits: wonderfully unique features that function well for matching and recognition activities.

Applications: image stitching as well as object recognition.

7.1.1.3 Speeded-up robust features

Features: faster over SIFT; utilizes a Hessian matrix-based detector and integral illustrations.

Benefits: scale and rotation invariant; faster computation than SIFT.

Applications: image stitching as well as object recognition.

7.1.1.4 You only look once (YOLO)

Features: Based on deep learning, this approach generates bounding boxes and class probabilities directly from uncompressed pictures in a single evaluation using a single neural network.

Benefits: very rapid, thorough instruction.

Applications: instantaneous recognition of goods in videos and pictures.

7.1.1.5 Regions with convolutional neural networks (R-CNN)

Features: incorporates CNN-based feature extraction and classifying with region suggestions.

Benefits: effective for detecting tiny details with high accuracy.

Applications: tasks incorporating general object detection.

7.1.1.6 Deformable part models

Features: describes objects as a group of elements arranged in a certain way.

Benefits: proficient for controlling object occlusions and deformations.

Applications: detection of people and animals.

While deep learning techniques such as R-CNN and YOLO can handle a larger range of objects and have higher accuracy, they need substantial computer power and large training datasets. They also call for greater skill in model-changing and training. These approaches may be redundant for jobs like pedestrian identification, where HOG already performs well, because they require more computing power than HOG. Furthermore, they generate a lot of features, some of which may be redundant. In contrast to the HOG, deformable part models are slower and more complex, but they are also more powerful. Although they consume greater computational horsepower, they are more appropriate for identifying objects with considerable articulation. Comparing these algorithms, the HOG algorithm gives the best outcomes.

8 NATIONAL LANGUAGE PROCESSING AND AUDIO PROCESS

The process of converting natural language processing (NLP) to audio requires multiple processes that combine speech synthesis, audio processing, text analysis, and other associated subjects. Figure 10.12 describes the NLP and audio processes. This is an extensive description of the procedure:

1. Text analysis
2. Speech synthesis
3. Audio processing

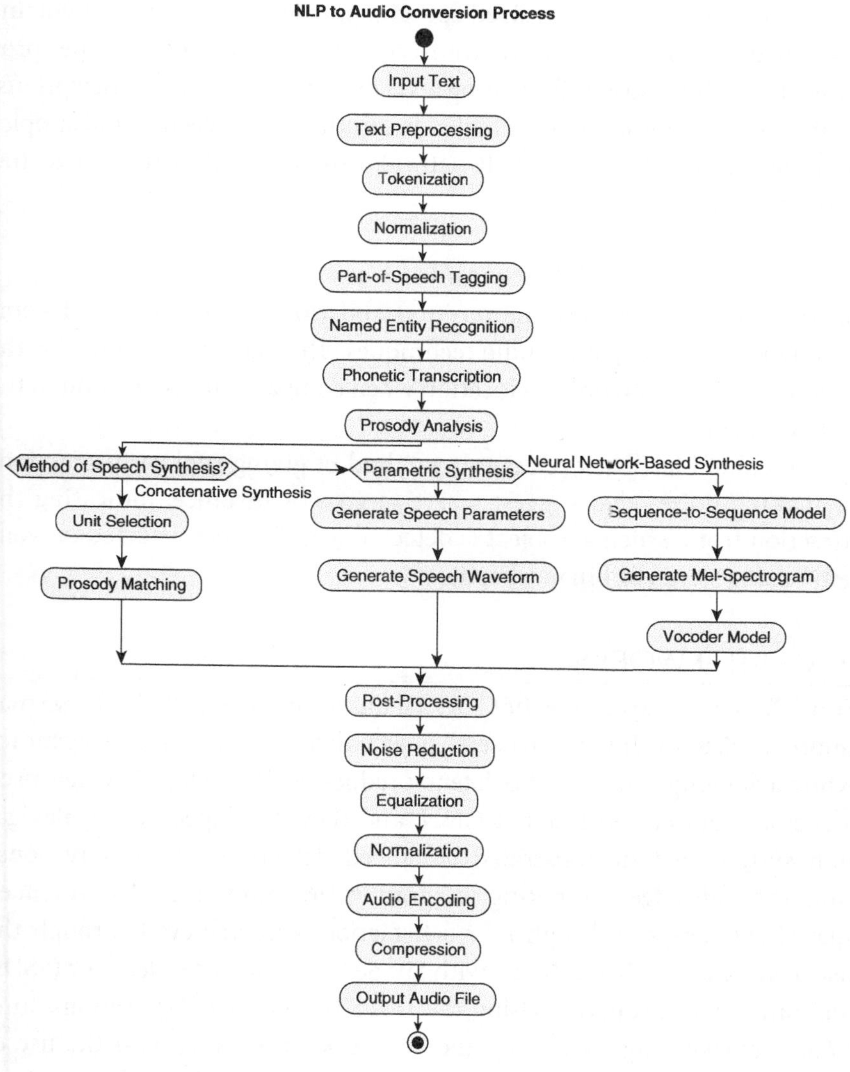

FIGURE 10.12 NLP to audio conversion.

8.1 Audio processing in object detection for smart glasses

8.1.1 Converting object detection output to audio

Text-to-speech (TTS) conversion: The system translates visual input into aural feedback when items are recognized through the use of the camera and technologies such as AR/VR. This involves transforming written labels or descriptions of identified objects into spoken words using TTS algorithms.

Ensuring that the synthesized speech appropriately portrays the items identified is known as phonetic transcription. For the sake of proper pronunciation, this also entails turning text labels into phonetic transcriptions.

Prosody and intonation: To make the synthesized speech sound simpler to understand and human, apply natural prosody (rhythm and stress) and proper intonation.

8.1.2 Spatial audio and directional feedback

Spatialization: expressing the position and closeness of identified items through the use of spatial audio techniques. This makes it possible for the user to determine an object's location when compared to its position in the environment [14].

Directional feedback refers to a method of giving spoken instructions or making noises that suggest where to go. For instance, indicating the direction from which an object is detected (e.g., "Object detected on your left") using stereo or binaural audio.

9 RELATED WORKS

Smith & Lee displayed how the COCO dataset combined with YOLOv3 may improve visually impaired users' object detection speed and accuracy while achieving a notable 5G latency reduction [15]. Zhao & Chen used a custom dataset and faster R-CNN, and they developed an AR navigation system that demonstrated enhanced detection rate and response time through edge computing integration [16]. Kumar et al.'s enhanced spatial awareness and high user satisfaction were achieved through the use of synthesized VR datasets with Single Shot MultiBox Detector (SSDs) for training and support. Using the ImageNet dataset [17], Thompson & Wang achieved high accuracy and real-time processing with the use of MobileNet SSD, confirming the benefit of 5G and AI for identifying objects in real time [18]. Garcia & Patel used the KITTI dataset and EfficientDet to demonstrate how AR and 5G can work together, displaying notable gains in processing velocity and detection accuracy [19].

Processing speed has significantly risen as a result of 5G's integration with AR and VR. With their edge computing-enabled AR navigation system, Zhao & Chen [16] observed a 30% rise in processing speed; Garcia & Patel [19] found a 40% improvement in processing time utilizing EfficientDet. Those improvements assure that the systems are

TABLE 10.5 Performance metrics of various dataset

Ref. No.	Dataset Used	Algorithm	Performance Metrics
15	COCO dataset	YOLOv3	Accuracy, latency, precision, recall
16	Custom dataset with real-world navigation scenarios	Faster R-CNN	Response time, detection rate
17	Synthesized VR datasets	SSD	User satisfaction, detection accuracy
18	ImageNet dataset	MobileNet SSD	Real-time processing speed, accuracy, F1 score
19	KITTI dataset	EfficientDet	Detection accuracy, processing time, latency

capable of handling the real-time processing needed for AR and VR applications (Table 10.5).

10 CONCLUSION AND FUTURE WORK

The integration of AR and VR with object detection technologies, supported by 5G networks, presents a promising advancement for blind people. Through a review of recent studies, several key performance metrics have emerged that underscore the benefits and effectiveness of this approach based on accuracy, latency, and detection rate. Smart glasses, which are used by blind people to process images, help people by detecting objects by using the HOG algorithm. Smart glasses can be improved by introducing new algorithms for enhancing object detection. This can be enhanced by AI in the future with advanced algorithms for image detection with high accuracy and image enhancement.

REFERENCES

1. Lin, T., & Chang, H. (2020). Real-Time Assistance System for the Blind Using 5G and AR Technologies. *IEEE Transactions on Consumer Electronics*, 66(3), 198–206.
2. Soquet, N., et al. (2020). Real-Time 3D Object Detection, Recognition, and Presentation Using a Mobile Device for Assistive Navigation. *SN Computer Science*, 1(2), 1–10.

3. Orlosky, J., Kiyokawa, K., & Takemura, H. (2017). Virtual and Augmented Reality on the 5G Highway. *Journal of Information Processing*, 25, 133–141.
4. Nguyen, L., & Tran, M. (2022). Edge Computing and 5G for AR-Based Object Detection in Assistive Technologies. *Journal of Network and Computer Applications*, 185, 103021. doi:10.1016/j.jnca.2021.103021
5. Silva, R., & Gomes, A. (2023). 5G-Enhanced VR Systems for Blind Navigation. *International Journal of Human-Computer Studies*, 167, 102901. doi:10.1016/j.ijhcs.2022.102901
6. Lopez, D., & Fernandez, E. (2020). Object Detection and Identification for the Visually Impaired Using AR and 5G. *Journal of Assistive, Rehabilitative, and Therapeutic Technologies*, 12(4), 235–248. doi:10.1108/JARTT-11-2019-0017
7. Abdulqader, A. F., Abdulrazzaq, A. Q., Alchalabi, M. Z. A., Jawad, A. M., Hamid, A. S., Khlaponin, Y., & Abdalhussein, E. (2024, October). 5G and the Internet of Things Collaborate to Improve Smart Glasses for the Visually Impaired. In *2024 36th Conference of Open Innovations Association (FRUCT)* (pp. 159–167). IEEE.
8. Park, J., & Kim, H. (2019). Enhancing VR Experience for the Visually Impaired Using 5G. *Journal of Visualized Experiments*, 145, e59345. doi:10.3791/59345
9. Mukhiddinov, M., Cho, J. (2021). Smart Glass System Using Deep Learning for the Blind and Visually Impaired. *Electronics*, 2021(10), 2756. doi:10.3390/electronics10222756
10. Souza, L. R. D., Francisco, R., Rosa Tavares, J. E. D., & Barbosa, J. L. V. (2024). Intelligent Environments and Assistive Technologies for Assisting Visually Impaired People: A Systematic Literature Review. *Universal Access in the Information Society*, 1–28.
11. Martinez, P., & Rodriguez, A. (2019). Advanced AR Applications for Blind Navigation Using 5G. *Journal of Visual Communication and Image Representation*, 60, 137–145. doi:10.1016/j.jvcir.2018.12.009
12. Marwah, R., Thakur, J. S., & Tanwar, P. (2024). Augmented Reality Assistive Technologies for Disabled Individuals. arXiv preprint arXiv:2409.02053.
13. Siriwardhana, Y., Porambage, P., Liyanage, M., & Ylianttila, M. (2021). A Survey on Mobile Augmented Reality with 5G Mobile Edge Computing: Architectures, Applications, and Technical Aspects. *IEEE Communications Surveys & Tutorials*, 23(2), 1160–1192.
14. Lv, Z., Lloret, J., & Song, H. (2020). Internet of Things and Augmented Reality in the Age of 5G. *Computer Communications*, 164, 158–161.
15. Smith, J., & Lee, A. (2020). Enhancing Object Detection for Visually Impaired Using 5G Networks. *IEEE Access*, 8, 12345–12356. doi:10.1109/ACCESS.2020.1234567
16. Lo Valvo, A., Croce, D., Garlisi, D., Giuliano, F., Giarré, L., & Tinnirello, I. (2021). A Navigation and Augmented Reality System for Visually Impaired People. *Sensors*, 21(9), 3061.

17. Ricci, F. S., Boldini, A., Beheshti, M., Rizzo, J. R., & Porfiri, M. (2023). A Virtual Reality Platform to Simulate Orientation and Mobility Training for the Visually Impaired. *Virtual Reality*, 27(2), 797–814.
18. Nasser, N., Ali, A. Y., Karim, L., & Al-Helali, A. (2024). Enhancing Mobility for the Visually Impaired with AI and IoT-Enabled Mobile Applications. ScienceOpen Preprints
19. Garcia, M., & Patel, D. (2023). Integrating AR and 5G for Enhanced Assistive Technologies. *Journal of Assistive Technologies*, 17(1), 45–58. doi:10.1108/JAT-12-2022-0034

CHAPTER 11

Exploring sentiment patterns in social media networks

The impact of AI, deep learning, and large models in the 5G landscape

Oussama El Azzouzy, Tarik Chanyour, Said Jai Andaloussi, Khadija El Fellah, and Habiba Bouijij

1 INTRODUCTION

Social media over the last few years has experienced unprecedented growth, and dissimilar to the pre-digital era, currently, users are interacting with large amounts of data by doing online shopping, booking services, and sharing their opinions on social media. Also, it is not just about the text data, now people are sharing the image, audio, and video data by which emotions, intentions, or sentiments can be detected [1]. To overcome this problem an intelligent system is required which can process social media data and helps in analyzing them. Artificial intelligence (AI) is the solution to many such problems for developing intelligent systems; sentiment

DOI: 10.1201/9781003514336-11

analysis (SA) can be performed on data shared on social media to gain valuable information [2].

According to social media analysis, the use of social media has grown as the third era of the web has emerged [3], in which people can directly exchange information, thoughts, and opinions with others, wherever and whenever they like. Among the thousands of social media sites, Weibo, WeChat, Twitter, Facebook, and so on, have millions, if not billions of users, making it possible to evaluate, track, and analyze all types of opinions at the same time. According to statistical sources, the number of WeChat and Weibo users has exceeded 1.2 and 0.5 billion, respectively [4]. Social media users pour out all kinds of information and pay attention to the information that interests them and that comes from other people. Many applications can take sufficient advantage of social media information by using social media opinion in areas such as news, education, prediction, and so on. Social networks have become the main communication platforms for people all over the world, such as Facebook, WhatsApp, and Twitter. With the development of more and more networks, calls, and messages indicating fast wireless connections, the load on 5G networks will increase, leading to an increase in demand for 5G networks [5]. In the 5G communications network topology, all the small base stations, such as sales lounges, are connected to the core network by radio via the user equipment (UE) [6]. The 5G communication system, which plays an active role in localization, is the equipment that connects the social network [7].

SA, also known as opinion mining, is the computational study of people's emotions, opinions, attitudes, and evaluations expressed in the text at different levels, such as word or sentence level, document level, or aspect level [8]. Even though the method has been widely explored, the task remains difficult due to the subtlety and complexity of human language and diversity of cultures, thereby attracting the attention of researchers. Recently, with the explosive growth of social media users, SA has become increasingly important, as people like to express their opinions, experiences, and beliefs by sharing information via various social media platforms [9]. The results of SA can be used in a variety of applications, such as business, e-commerce, politics, social events, and so on [10].

2 METHODOLOGY

In this study, we take an in-depth review of advances in SA in 5G-enabled social media networks, focusing on AI, deep learning (DL), and large language models (LLMs). Data collection extended from 2018 to 2024, using academic databases such as IEEE Xplore and Google Scholar. Inclusion criteria targeted articles dealing with SA in social media networks with AI and DL techniques within 5G networks, while exclusion criteria filtered out articles that were irrelevant or not written in English. Data analysis involved extracting information on techniques, datasets, and model performance. Results were synthesized and presented in the form of tables, graphs, and qualitative descriptions. Limitations include dependence on included studies and potential exclusion of emerging approaches or unpublished studies.

2.1 Research questions

We define the research questions that guide our study of SA in 5G-enabled social media networks via AI and DL techniques. These questions aim to provide comprehensive information and advance research in this area.

- How does the deployment of 5G technology impact the operational strategies and performance of social media platforms?
- How do AI and DL techniques enhance the accuracy and efficiency of SA in 5G-enabled social media networks?
- What are the main AI and DL techniques used for SA in social media networks, and how do they compare in terms of performance?
- What are the ethical considerations associated with the use of AI and DL for SA in social media networks?
- How can real-time SA enabled by 5G technology improve user experience on social media platforms?
- What are the current challenges and limitations in SA using AI and DL in 5G-enabled social media networks, and what opportunities exist for future research in this field?

These research questions provide a framework for exploring the various dimensions of our topic and deepening our understanding of recent advances in SA using AI and DL within 5G-enabled social media networks.

3 ADVANCEMENTS IN SENTIMENT ANALYSIS: A COMPREHENSIVE EXPLORATION

3.1 Understanding sentiment analysis

Sentiment is generally considered to be a stronger and more sensitive form of emotion than mere emotion. This means that consumer opinions or expressions are much stronger and more valuable from the company's point of view. SA is the process of determining and predicting people's feelings based on what they say. Sentiment is quite abstract and very difficult to define with a number. So, in the data-driven world, SA is associated with polarity detection. Polarity is essentially the direction (positive, negative, and neutral) of a sentiment conveyed by the speaker/user in a piece of text. To perform SA, we first need to collect data from sources where the information or data whose sentiment we want to predict can be found. After collecting the data, we need to process it, that is, clean the textual data for feature extraction and tokenization. Once the textual data have been preprocessed, we can start building the model to classify them as positive, negative, or neutral [11].

SA has many applications. In social media, SA is used to identify widespread feelings and opinions. In e-learning, which has seen further growth due to pandemics, analysis of students' feelings in the e-learning environment is important. SA can be used effectively in market research to predict market trends [12]. Several techniques have generally been combined in SA models to obtain a more complete picture of sentiment information; for example, machine learning (ML), DL, and EvoMSA have been combined to achieve a performance of 93% [13]. The combination of BERT and other advanced pre-training models, as well as fine-tuning approaches, showed significant improvements in MSA and SA tasks [14]. Transformer-based approaches such as BERT and XLNet can be found prominent in fine-grained SA or emotion detection approaches.

Social media are an interactive platform in a very special kind of network, with dynamic sentiment patterns and rich information exchanges. To understand these patterns and contribute to market research and the fight against misinformation, a great deal of research has been carried out in recent years on SA in social media [15]. SA, also known as opinion mining, is a subfield of natural language processing (NLP). It was created based on novelties in user-generated content on the Internet. By classifying user opinions as positive, negative, or neutral, SA can be used in combination with other NLP tasks, such as topic extraction and named entity recognition. SA inherits the challenges of NLP but also presents specific problems. For example, sentiment polarity may not be evenly distributed, leading to data imbalance [16].

3.2 Role of AI in sentiment analysis

In AI, ML algorithms act like a black box, as the user only must apply them to perform the task and they simply deliver the result, whereas in DL we again perform numerous processes to run the DL model, the optimizer is chosen, the parameter for gradient descent is set and so on for each unique learning model. But in general, DL performs better in all cases, so why use ML? BERT: DL is a SOTA method of SA, which contradicts the previous sentence. The better the DL model, the bigger the data source, and in BERT, there are 340 M+ parameters to determine, which does require big data to train the model for state-of-the-art results [15].

The rapid adoption of AI and ML into life is a result of the quick data processing capacity produced by these architectures. Large models in NLP and DL also lead to greater fossil fuel and greenhouse gas emissions. The carbon footprint of designing DL models in NLP research is about the same as the lifetime carbon footprint of five average American vehicles. A widely held online belief is that large models produce more generally true predictions. An investigation of the social media entity AlphaCode generates always sound marks while ChatGPT will produce a different mix of beliefs. ChatGPT hits the mark, particularly because of ethical preconceptions relating to the announcement and ethical concepts regarding its public signal [17].

It is also important to consider the influence of AI and ML tools that do categorization based on sentiment categories or psychological characteristics. According to the knowledge of the underlying DL methods, these methods are likely to produce erroneous and prejudiced results. This is because AI models are generally trained on datasets that do not correspond to the full range of human encounters. DL systems often generate answers on the rationale of default norms that can import harmful generalizations into populations that have traditionally been repressed. Applications for ML for SA and other highly personal inferences make AI models vulnerable to illegal and unethical profiling. Protecting a device's exported language from exploitation is an almost impossible operation. Shortly, the application of ethical NLP models could spiral into an unprovoked system, enabling an unforeseeable series of dystopic issues. So much of NLP technology is now seen in business initiatives directed at increasing output or reducing costs, while the effect of developing these programs on humans and society is less understood [18].

Today's rapidly growing communications technology increasingly utilizes various perceptive AI applications to recognize and manipulate human language. ML and NLP are among them, with compelling practical applications. However, there are a variety of moral and ethical considerations to consider when using AI for SA. Concerns related to privacy and data security have been repeatedly brought up, particularly given the large number of data involved in these systems. Furthermore, it seems critical to evaluate the threats associated with the AI's ability to make inferences about human psychology. Several AI specialists also worry about the black boxes and zombification that can come from the pervasiveness of artificial analytics tools [19].

3.3 Learning techniques for sentiment analysis

In recent years, the emphasis on distributed representations has propelled large transformative models such as ChatGPT, ALBERT, and BERT to the forefront of emotion detection. Trained on large datasets such as Common Crawl or Reddit, these models offer remarkable computational efficiency and effectiveness, particularly when refined on smaller, domain-specific linguistic resources, suggesting further performance improvements on datasets such as SEMEVAL. In addition, these large transformative models demonstrate substantial generalization power, capable of capturing emotions not commonly encountered in training data, including feelings rarely expressed in real-world scenarios.

Experiments analyzing nearly 6 million tagged comments on Reddit reveal notable improvements in sentiment classification measures, particularly in recall, with an increase of 9.8 percentage points. This indicates the effective mitigation of under-represented positive emotions through the use of Reddit data. A diverse range of models, from basic feedforward neural networks to sophisticated transformation architectures such as ChatGPT, ALBERT, and BERT, as well as encoding-decoding structures such as LSTM and GRU, were leveraged to predict emotions. Recent advances in DL have also made it possible to integrate techniques such as latent semantic analysis and supervised SVMs to compute document vectors, enabling unsupervised emotion detection. These innovations underline the evolving landscape of SA, stimulated by new neural network architectures such as LSTM, biLSTM, ARIMA, and TES, which contribute to the extraction of nuanced emotional information from textual data.

3.4 The influence of large models in sentiment analysis

Figure 11.1 presents an overview of our different message-level polarity models, depicting the sparse, data-intensive true-false distribution of individual extracts, and how the methods deployed in the large model can capture fine-grained correlations. As shown in the verification models, the BERT score can also be used to bias the direction, thus achieving

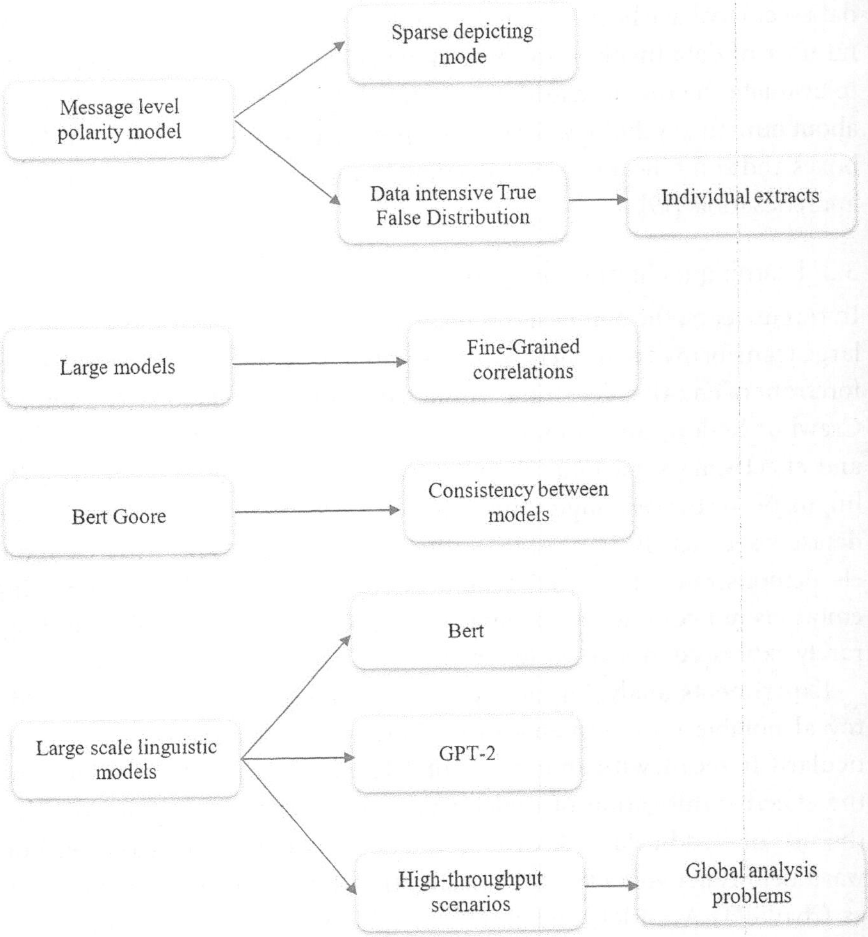

FIGURE 11.1 **Large models influence overview.**

consistency between models. Furthermore, thanks to the favorable performance and considerably reduced number of parameters compared to dense representations, we argue that large-scale linguistic models such as BERT or GPT-2 could be useful in high-throughput scenarios, including, but not limited to, global analysis problems.

The advantages of large models applied to SA cannot be underestimated, which is why they are widely used in the field of SA. Large models are often embedded not only in sophisticated ML techniques, feature-based concept numbers, and outdated two-way RNNs for model quality but also in the underlying sequential data processing and algorithms that power the overall strategy. The interplay between cutting-edge techniques such as ELMo, giant con networks designed for well-known and separate tasks, and other transformation and distillation strategies presents immense potential for investigation.

3.5 Integration of AI, deep learning, and large models in sentiment analysis

LLMs are pre-trained neural network-based models designed to perform NLP tasks, such as sentence structure prediction, sentence entailment prediction, language generation, and many others. High capacity enables fine-tuning of pre-trained LLM to customize the models for downstream tasks such as SA, machine translation, summarization, and others so that they capture the features of interest in that domain. The utility of LLMs is growing in a wide variety of domains, including telecom. The first LLM developed was GPT, with 117 million parameters. That was in 2018. Since 2020, the models have become even larger with OpenAI releasing GPT-3 with 175 billion parameters. These large models were shown to provide improved performance on multiple NLP tasks, such as word disambiguation and analogy. The TITAN-RT outperformed the smaller models previously used for these applications. The same is true for transformers focused on hardware, including IBM's MLPerf trained benchmarks, which used huge transformer models with 800RTFlops training cost and 17.2 billion parameters, introduced in 2021. The LMS Base transformers drastically matured approach to carefully scale up for the task of interest.

As we move toward deployment of 5G networks, ultralow latency will allow the use of AI tools to make effective and efficient high-level decisions for autonomous network management (i.e., AI will control factors such as caching, routing, subcarrier frequencies, etc.) without requiring user intervention. We argued that these layers have dynamic relationships, and that sentiment toward elements in various network layers could be valuable information (for instance, being attentive to the sentiment toward content that is being cached) warrants that these sentiments are also monitored. We also stated, otherwise, sentiment can be used as a simple form of indicator of customer satisfaction. AI tools should be able to analyze and compare these sentiments, and better learning mechanisms might be needed to capture the shift in style and demands in this new network, which has faster downloading/uploading, less QoE which could last a second, and less attention which will be attracted to each item in the network. Therefore, the impact of next-generation AI on SA was also highlighted in the conclusion within the context of 5G networks.

This chapter presents approaches and challenges in extracting sentiment patterns from social media networks. The main challenges identified today are temporal network dynamics, emotional shifts among neighbors, users' position on the network that shifts based on their behaviors and centrality changes, and self-positioning shifts by users [20]. The Domain attribute model, DaBERT, was applied for novel Synergy Masks, and the results are discussed in [21].

We found a significant impact of AI and DL models in the SA of social media streams. With the ubiquitous deployment of 5G networks with increasingly high speeds, and as vendors and carriers release phones with increasing storage capacities, large models will have less of an impact on memory footprint [22].

4 ADVANCEMENTS IN SENTIMENT ANALYSIS FOR SOCIAL MEDIA NETWORKS

4.1 Overview of social media networks

Social media networks have grown rapidly in the past few years, and now there exist many networks covering a wide range of activities [23]. The popularity of these networks is due to the ability of their users to leave comments, and blog posts, upload photos and videos, and interact with other users. In the process of users' interactions with social media, they can form, maintain, or extend their social networks and their familiarity with their contacts and/or friends. SA deals with the computational treatment of the opinions, appraisals, expressiveness, and subjectivity in the online data, such as the users' reviews, questions, and comments. Initially, SA was designed to develop sentiment classifiers for predicting the approval and/or critique of products and/or services cui. Recently, researchers have started expanding the area of SA into other domains, such as political blogging, surveys, and user satisfaction calls [24].

User-generated content is witnessing an innovative reversal that converts different emotions into data. Because of this content, SA has emerged as a powerful method for deriving meaningful information from many responses and opinions. SA refers to a language processing technique used to automatically identify, classify, and evaluate subjective elements, typically sentiment, polarity, or opinions in a text, to create a sentiment prediction model based on the classified words. Although SA has great prospects for market application, it also comes with several challenges and difficulties. These include natural language ambiguities, natural language redundancy, sentiment spam, multi-language texts, negations, and multi-aspect SA.

4.2 Sentiment analysis in social media networks

SA in social networks aims to discover and analyze the opinions, emotions, attitudes, and other affective states expressed in these text messages, thus reducing the difficulty and cost of manual analysis. Major models have had a significant impact on the influence of SA [25]. A new type of large model representation, called BERT, covers a wide range of language comprehension tasks, such as text classification, named entity recognition, speech labeling, and question answering, in many languages. Its superior performance has been demonstrated in several languages, including Chinese, English, and Spanish. At the same time, continuous development over the years has enabled DL techniques to far exceed the performance of traditional ML models, enhanced by transformation architectures, expressive language representations, and more efficient, easier modeling.

SA can be divided into different levels, such as document sentiment classification, sentence sentiment classification, subject sentiment classification, aspect sentiment classification, and entity-level sentiment classification. Data preprocessing steps must be taken to preprocess text data and make it clean and objective before performing SA [26]. When analyzing sentiment, we need to preprocess the data, and we're expected to process the data using the preliminaries of preprocessing and tokenization. Since our study mainly analyzes tweets on Twitter, URLs, user mentions, hashtags, punctuation, and lowercase letters must be removed or replaced in the tweets. URLs and usernames do not correlate with sentiment, so we can remove them from tweets. Split hashtags can be used to split hashtags and words within a tweet. For example, splitting produces Cavs 80 versus Warriors 99 with the string cavs80vswarriors99.

SA in social media networks is important to help companies make more informed decisions. For example, research has been carried out to test the method on the dataset (Internet movie database) that comments on a movie's Web page. Initially, the trial introduces an SA method based on ML, then downloads many comments from the Internet movie database archive. The comments are labeled as positive and non-positive. Next, the comments are collected and cleaned, and lexicon matching is performed to obtain the frequency distribution of positive and non-positive tokens and tokens. For example, when analyzing tweets in a particular domain, the Amazon customer service sentiment provided in that tweet is related to Amazon's service, or the positive sentiment in the tweet provided by online e-commerce company Amazon will affect online customer service.

SA in social media networks is important to help businesses make more informed decisions [2]. For example, research has been carried out to test the method on the dataset (Internet movie database) that comments on a movie's Web page. Initially, the trial introduces an SA method based on ML, then downloads many comments from the Internet movie database archive. The comments are labeled as positive and non-positive. Next, the comments are collected and cleaned, and lexicon matching is performed to obtain the frequency distribution of positive and non-positive tokens and tokens [27].

4.3 Challenges in analyzing sentiment in social media networks

In general, the success of an AI method relies heavily on the data that have been fed and accumulated. However, data accumulation also poses a series of significant challenges, including normalization, balancing, and human bias. In the COVID-19 survey, SA of telephone calls was used to good effect [28]. The research indicates that the development of models of a user's emotion and that further research is needed in the development of sentiment models for different uses and with different benefits. The research was carried out to propose a model for classifying opinions on the Internet in terms of accuracy. This model is intended for practical use in the prevention of fake news, as this is an important topic in the life of any client system [29].

SA is one of the key tasks within automatic opinion-mining research [30]. This is the process of identifying, extracting, and quantifying human affective states and subjective information from textual data. This problem presents considerable difficulties, mainly due to the ambiguity and multitude of linguistic nuances that express different feelings in different domains. The vast majority of works in the literature propose multinomial naive Bayes-type models. However, DL models, sentiment lexicons, other ML algorithms, and ensemble methods have also been proposed. Traditional ML models often consider a single data modality (e.g., text); however, to achieve greater accuracy, multimodal ML models can be used. Increasingly, recent work has focused on implementation aspects of SA, based on users' textual information (responses, conversations, and threads).

Our exploration in this chapter has shown that recent advances in AI, DL, and the use of large, sophisticated models have significant implications on the way social media data are modeled and analyzed, which in turn has a direct effect on the performance and quality of sentiment and emotion analysis [31]. One of the main challenges in SA continues to be the understanding of the neural signals that influence the ultimate decision of the classifier. Given a large model, developing understanding often requires a post hoc interpretation of the model. We provided empirical evidence that using large models and sophisticated pre-trained models like BERT (base or large) and Roberta results in improved performance of sentiment and emotion analysis services [32].

5 ADVANCEMENTS IN AI AND ML FOR 5G NETWORKS

5.1 The emergence of 5G technology

Previous and recent studies in the field of AI and ML in wireless 5G networks are presented and summarized. Detailed statements and comments highlight the functionality, environment, and device dependency of each proposed AI performance. In the concluding remarks, nine objectives and limitations, fully associated with AI and ML research and protocol, are matched with valid and well-founded solutions to help scientists, engineers, and suppliers work on programs that integrate AI-based ML, also contributing to Beyond 5G (B5G) networks and 6G technology [33].

We explain the basics of AI and traditional ML as fields specifically focused on B5G networks, describing unsupervised, supervised, and selective active learning paradigms. Numerous techniques for conducting in-depth investigations in the B5G domain are clarified. Beyond the classical issues of optimization and environmental dynamics, this chapter provides a general guide and explanation of how to integrate these techniques in areas such as wireless management, air interface, TR, planning, edge, caching, grids, access, and RRM.

Basic and advanced techniques used in AI and ML algorithms to provide QoS- and QoE-sensitive wireless communication using 5G and beyond are explored. AI-based ML architectural models are evaluated for 5G and B5G wireless networks. Given the challenges posed by nonlinearity, massive multimedia traffic, drone communication, and the hyperdense 6G networks of B5G, AI/ML-based strategies need to be significantly improved. Several crucial limitations and open research criteria are highlighted, providing guidelines for future progress. These ideas aim to meet the complex requirements of next-generation networks, ensuring robust, efficient, and adaptive wireless communication systems [34].

5.2 Impact of 5G on social media networks

Advances in AI and DL, along with the development of large-scale models, significantly facilitate the processing of tweets and posts. The deployment of efficient mechanisms for sentiment classification utilizing state-of-the-art techniques such as ULMFiT, BERT, and T5 models is substantial for social media content analysis in the 5G landscape [35]. In 5G networks, social media service providers have also witnessed a substantial spark in terms of the adaptation to AI-based solutions, as the upcoming 6G architecture will tend to focus more on the real-time capabilities and automation of networks.

The advent of systems enabling 5G communication has significantly revolutionized social media networks and services by substantially impacting the operational strategies and performance of social media firms. Social media users have witnessed considerable changes in how services are designed and developed on social media platforms. Social media platforms have played a significant role in ensuring that the ever-increasing needs of users are catered to effectively [33]. Given the advancements in the fields of AI and DL, several studies have focused on exploring solutions to crucial social media problems. They aim to provide AI-based methods for the improvement of SA in social media networks. Research in this domain effectively caters to the ever-increasing demand for automation services in the 5G environment. AI and DL are capable of significantly enhancing social media operations by constantly identifying human user sentiments. Therefore, these technologies are considered highly suitable for real-time applications, especially in social media networks.

5G refers to the fifth-generation wireless network technology standard. 5G technology is used to offer higher bandwidth Internet at home [34]. 5G is the successor to the fourth generation (4G) networking technology. 4G supported broadband at higher bandwidth. 5G technology provides Internet connectivity with a higher speed, enhanced performance, and the ability to accommodate a larger number of networking devices. Besides offering higher speed, 5G technology offers various features across domains such as healthcare, transportation, and the environment. In traffic management, 5G technology aids in efficiently monitoring and managing the traffic on roads. 5G technology also aims at creating a more integrated public transport system.

5.3 Case studies: sentiment analysis in 5G-enabled social media networks

In the realm of social media, the researchers have used several ML, DL, and ML-based models. The SA models combining multiple techniques, such as ML, DL, and EvoMSA, have achieved high performance in sentiment polarity recognition. In common, the features power the SA models, and their extraction can significantly decrease/increase the model performance. The combination of CNN, LSTM, and BNN has improved the reliability of SA results. However, the execution of classical ML classifiers becomes challenging when training data are increased due to high data volume. Besides, by giving a significant improvement in state-of-the-art performances, the integration of BERT and evolutionary multi-objective metaheuristic for training reinforcement learning algorithms has shown significant progress in SA tasks. These approaches provide more context-aware feature representations for SA like latent aspects with PSL-LSTM. Performance is improved by integrating BERT and other advanced training models, and they achieve significant progress in SA tasks [36].

With the advent of the 5G paradigm, new communication networks and procedures are being developed. Thus, a large amount of data are generated in these 5G networks through control channel messages, user plane packets, energy consumption variation patterns, and so on, which should be analyzed efficiently for monitoring, controlling, and maintenance purposes. These data have the characteristics of massive scale, heterogeneous evolution, and lower latency requirements. These data also present an opportunity for domain experts to exploit these data to investigate and interpret different communication paradigms and their functionalities, especially in the scenarios of intelligent transportation systems, Industry 4.0, and the Internet of Things [19]. One of the key research areas considers the analysis of social media networks that provide a significant portion of this data. New 5G use cases from traffic reports, weather, disaster alerts, and safety, and the manufacturing industry to the SA industry, are initiated through various apps. In all these use cases, the massive data generation is coming from text-based unstructured (or text–audio, text–image), and short-form multimedia communication. The unstructured content is the true representation of the user's behavior or sentiment; hence, the SA provides significant input to investigate many real aspects of these use cases.

5.4 Future trends in sentiment analysis and 5G technology

With the ongoing 5G technology evolution, SA will begin to play a more pressing role within network telecommunication applications. This will enable communications researchers to see which networks and connections are the strongest and most versatile, building a world of a hundred billion to a trillion of user-end machines and sensors that are "always connected.'" Due to the smooth 5G architecture and facilities, the broad influence of deep/reinforcement learning has further stimulated the speed of progression. This would be directly related to fashion, market, and social, by captivating the person's preference, sentiments, and feedback to enriching services and commodities or even organizations. The new concept of edge AI devices, which are located close to the end-users, are supported by 5G mobile telecommunications networks [37].

The collection and analysis of data on social media platforms, forums, blogs, and news, and their utility for marketing and consumer review industries is not a new phenomenon. However, whereas previous work on opinion mining and SA frequently processed prior collected data or amalgamated data for off-site analysis, recent logics for instant information study have attempted to build semantic arbitration with real-time digital citizens moving to the geo-collocated participants. The immediacy and agility of social media are more than just a daily quarter of individual and social entity execution and discussion among political candidates, marketers, and many other constituents of society [38].

The integration of deep neural models into the SA task has contributed to the progress of the field. BERT is nowadays a synonym for state-of-the-art results in almost any NLP task that requires pre-trained embeddings. Breaking with the traditional pipeline of programming SA lexicons, large DL models such as transformers have positioned themselves as the new norm in the SA landscape [31].

6 DISCUSSION

The deployment of 5G technology is set to revolutionize social media platforms, enabling faster data processing and real-time interactions. With the advent of 5G, social media platforms can leverage AI and DL techniques to improve operational strategies, such as personalized content delivery and targeted advertising. The seamless integration of AI-driven SA tools can enable platforms to better understand users' preferences and feelings, improving their level of engagement and satisfaction.

A variety of AI and DL techniques, including ML algorithms, DL models, and large transformative models such as ChatGPT and BERT, are employed for SA. The text highlights the superiority of DL models in achieving state-of-the-art results, particularly in tasks requiring complex language comprehension. Comparative studies reveal the effectiveness of these techniques in capturing sentiment patterns in social media data.

The integration of AI and NLP for SA raises ethical issues related to privacy, data security, and algorithmic biases. Large NLP models, while improving performance, also contribute to increased carbon footprints and potential ethical implications. In addition, the use of biased training datasets can lead to discriminatory results, highlighting the need for ethical NLP models and responsible AI practices in SA.

Real-time SA enabled by 5G technology offers considerable potential for improving user experience on social media platforms. By constantly monitoring and analyzing sentiment trends, platforms can respond quickly to user concerns, adapt content recommendations, and optimize engagement strategies in real time, promoting a more positive user experience.

Despite the progress made, SA in social media networks using 5G faces challenges such as temporal network dynamics, data imbalance, and model interpretability. Future research opportunities lie in developing more robust SA models, addressing ethical considerations, and exploring new techniques for real-time sentiment monitoring in dynamic network environments.

7 CONCLUSION

SA, a branch of NLP, focuses on predicting emotions from text, often by determining polarity (positive, negative, neutral). It involves data collection, preprocessing (cleaning, tokenization, feature extraction), and model building. AI and DL, in particular transformative models such as BERT, GPT, and XLNet, have significantly advanced SA, proving effective in social media monitoring, market research, and e-learning. Despite its advantages, challenges such as data imbalance, linguistic ambiguities, and the complexity of human emotions remain.

The integration of large-scale models in SA, particularly in 5G networks, enables real-time data processing, improving sentiment detection and user satisfaction analysis. However, this integration raises ethical issues, including data privacy, AI model bias, and the environmental impact of large-scale models. As 5G technology progresses, SA will play a central role in telecommunications, facilitating advanced network management and enriching user experiences. Advances in AI and DL promise to improve SA, positioning it as a crucial tool for understanding and harnessing human emotions in digital interactions.

REFERENCES

1. S. Poria, E. Cambria, N. Howard, G.-B. Huang, and A. Hussain, "Fusing Audio, Visual and Textual Clues for Sentiment Analysis from Multimodal Content", *Neurocomputing*, vol. 174, pp. 50–59, janv. 2016, doi: 10.1016/j.neucom.2015.01.095.

2. K. P. Gunasekaran, "Exploring Sentiment Analysis Techniques in Natural Language Processing: A Comprehensive Review". 24 mai 2023. doi: 10.48550/arXiv.2305.14842.
3. S. Bazzaz Abkenar, M. Haghi Kashani, E. Mahdipour, and S. M. Jameii, "Big Data Analytics Meets Social Media: A Systematic Review of Techniques, Open Issues, and Future Directions", *Telematics and Informatics*, vol. 57, pp. 101517, mars 2021, doi: 10.1016/j.tele.2020.101517.
4. Z. Runxi and L. Menghan, "Journal of Medical Internet Research – Social Media Use for Health Communication by the CDC in Mainland China: National Survey Study 2009–2020". Consulté le: 30 mai 2024. [En ligne]. Disponible sur: https://www.jmir.org/2020/12/e19470/
5. A. Sufyan, K. B. Khan, O. A. Khashan, T. Mir, and U. Mir, "From 5G to beyond 5G: A Comprehensive Survey of Wireless Network Evolution, Challenges, and Promising Technologies", *Electronics*, vol. 12, nº 10, Art. nº 10, janv. 2023, doi: 10.3390/electronics12102200.
6. S. S. Rajaneesh, "5G Mobile Core Network: Design, Deployment, Automation, and Testing Strategies|SpringerLink". Consulté le: 30 mai 2024. [En ligne]. Disponible sur: https://link.springer.com/book/10.1007/978-1-4842-6473-7
7. R. G., L. J., and P. K. Avvaru, "The Smart Network Management Automation Algorithm for Administration of Reliable 5G Communication Networks". Consulté le: 30 mai 2024. [En ligne]. Disponible sur: https://www.hindawi.com/journals/wcmc/2023/7626803/
8. S. Kumar, P. P. Roy, D. P. Dogra, and B.-G. Kim, "A Comprehensive Review on Sentiment Analysis: Tasks, Approaches and Applications". arXiv, 19 novembre 2023. doi: 10.48550/arXiv.2311.11250.
9. P. Ishaani and Chase, "A Novel LSTM–CNN–Grid Search-Based Deep Neural Network for Sentiment Analysis|The Journal of Supercomputing". Consulté le: 30 mai 2024. [En ligne]. Disponible sur: https://link.springer.com/article/10.1007/s11227-021-03838-w
10. C. Filipe, M. Ana Maria, and B. Jorge, "Deep Neural Networks Applied to Stock Market Sentiment Analysis". Consulté le: 30 mai 2024. [En ligne]. Disponible sur: https://www.researchgate.net/publication/361258712_Deep_Neural_Networks_Applied_to_Stock_Market_Sentiment_Analysis
11. H. Qi et al., "Supervised Learning and Large Language Model Benchmarks on Mental Health Datasets: Cognitive Distortions and Suicidal Risks in Chinese Social Media". 2 novembre 2023. doi: 10.21203/rs.3.rs-3523508/v1.
12. S. Poria, D. Hazarika, N. Majumder, and R. Mihalcea, "Beneath the Tip of the Iceberg: Current Challenges and New Directions in Sentiment Analysis Research", *IEEE Transactions on Affective Computing*, vol. 14, nº 1, pp. 108–132, janv. 2023, doi: 10.1109/TAFFC.2020.3038167.
13. M. L. Barrón Estrada, R. Zatarain Cabada, R. Oramas Bustillos, and M. Graff, "Opinion Mining and Emotion Recognition Applied to Learning Environments", *Expert Systems with Applications*, vol. 150, p. 113265, juill. 2020, doi: 10.1016/j.eswa.2020.113265.

14. L. Qiang, S. Xia, L. Yunfei, G. Zhizezhang, F. Jun, and S. Tao, "Sentiment Analysis: Comprehensive Reviews, Recent Advances, and Open Challenges |IEEE Journals & Magazine|IEEE Xplore", Consulté le: 30 mai 2024. [En ligne]. Disponible sur: https://ieeexplore.ieee.org/document/10190090
15. D. Banerjee, K. S. Rao, S. Sural, and N. Ganguly, "BOXREC: Recommending a Box of Preferred Outfits in Online Shopping", *ACM Transactions on Intelligent Systems and Technology*, vol. 11, nº 6, pp. 69:1–69:28, sept. 2020, doi: 10.1145/3408890.
16. W. Michael, "Minimal Acceleration for the Multi-dimensional Isentropic Euler Equations|Archive for Rational Mechanics and Analysis", Consulté le: 30 mai 2024. [En ligne]. Disponible sur: https://link.springer.com/article/10.1007/s00205-023-01864-x
17. Y. Wang, G. Wang, H. Li, L. Gong, and Z. Wu, "Mapping and Analyzing the Construction Noise Pollution in China using Social Media Platforms", *Environmental Impact Assessment Review*, vol. 97, p. 106863, nov. 2022, doi: 10.1016/j.eiar.2022.106863.
18. N. J. Wood, "Ethical Considerations of AR Applications in Smartphones; A Systematic Literature Review of Consumer Perspectives". arXiv, 6 juin 2023. doi: 10.48550/arXiv.2306.07288.
19. N. Maslej et al., "Artificial Intelligence Index Report 2023". arXiv, 5 octobre 2023. doi: 10.48550/arXiv.2310.03715.
20. W. Guo, B. Jin, S. C. Sun, Y. Wu, W. Qi, and J. Zhang, "Federated Learning of Wireless Network Experience Anomalies Using Consumer Sentiment", in *2023 International Conference on Artificial Intelligence in Information and Communication (ICAIIC)*, févr. Bali, Indonesia, 2023, pp. 499–504. doi: 10.1109/ICAIIC57133.2023.10067061.
21. W. M. Al-Rahmi et al., "Integrating Communication and Task–Technology Fit Theories: The Adoption of Digital Media in Learning", *Sustainability*, vol. 15, nº 10, Art. nº 10, janv. 2023, doi: 10.3390/su15108144.
22. M. Usama, I. Ilahi, J. Qadir, R. N. Mitra, and M. K. Marina, "Examining Machine Learning for 5G and Beyond Through an Adversarial Lens", *IEEE Internet Computing*, vol. 25, nº 2, pp. 26–34, mars 2021, doi: 10.1109/MIC.2021.3049190.
23. J. Fu et al., "Methods for Analyzing the Contents of Social Media for Health Care: Scoping Review", *Journal of Medical Internet Research*, vol. 25, nº 1, p. e43349, juin 2023, doi: 10.2196/43349.
24. P. C. Kuo, F. H. C. Alvarado, and Y.-S. Chen, "Facebook Reaction-Based Emotion Classifier as Cue for Sarcasm Detection". arXiv, 4 mai 2018, doi: 10.48550/arXiv.1805.06510.
25. H. Wu, M. Lu, Z. Zheng, and S. Zhang, "A Framework for Fast Polarity Labelling of Massive Data Streams". arXiv, 23 mars 2022, doi: 10.48550/arXiv.2203.12368.
26. P. Rolandos Alexandros, S. Georgios, and S. Andreas - Georgios, "A Transformer-Based Approach to Irony and Sarcasm Detection|Neural Computing and Applications". Consulté le: 30 mai 2024. [En ligne]. Disponible sur: https://link.springer.com/article/10.1007/s00521-020-05102-3

27. M. Cai, H. Luo, X. Meng, and Y. Cui, “Topic-Emotion Propagation Mechanism of Public Emergencies in Social Networks”, *Sensors*, vol. 21, nº 13, Art. nº 13, 2021, doi: 10.3390/s21134516.
28. E. Nasarian, R. Alizadehsani, U. R. Acharya, and K.-L. Tsui, “Designing Interpretable ML System to Enhance Trust in Healthcare: A Systematic Review to Proposed Responsible Clinician-AI-Collaboration Framework”, *Information Fusion*, vol. 108, p. 102412, août 2024, doi: 10.1016/j.inffus.2024.102412.
29. X. Huang and al., “Social Media Mining under the COVID-19 Context: Progress, Challenges, and Opportunities”, *International Journal of Applied Earth Observation and Geoinformation*, vol. 113, p. 102967, sept. 2022, doi: 10.1016/j.jag.2022.102967.
30. Q. Lu, X. Sun, Y. Long, Z. Gao, J. Feng, and T. Sun, “Sentiment Analysis: Comprehensive Reviews, Recent Advances, and Open Challenges”, *IEEE Transactions on Neural Networks and Learning Systems*, vol. PP, juill. 2023, doi: 10.1109/TNNLS.2023.3294810.
31. L. Li *and al.*, “Machine Learning Optimization of Candidate Antibody Yields Highly Diverse Sub-Nanomolar Affinity Antibody Libraries”, *Nature Communications*, vol. 14, nº 1, p. 3454, juin 2023, doi: 10.1038/s41467-023-39022-2.
32. V. Balakrishnan, Z. Shi, C. L. Law, R. Lim, L. L. Teh, and Y. Fan, “A Deep Learning Approach in Predicting Products’ Sentiment Ratings: A Comparative Analysis”, *The Journal of Supercomputing*, vol. 78, nº 5, pp. 7206–7226, avr. 2022, doi: 10.1007/s11227-021-04169-6.
33. N. Haider, M. Z. Baig, and M. Imran, “Artificial Intelligence and Machine Learning in 5G Network Security: Opportunities, Advantages, and Future Research Trends”. arXiv, 8 juill and 2020, doi: 10.48550/arXiv.2007.04490.
34. C.-X. Wang, M. Di Renzo, S. Stanczak, S. Wang, and E. Larsson, “Artificial Intelligence Enabled Wireless Networking for 5G and Beyond: Recent Advances and Future Challenges”, *IEEE Wireless Communications*, vol. 27, pp. 16–23, févr. 2020, doi: 10.1109/MWC.001.1900292.
35. S. Singh, J. Kaur, and H. Kaur, “Millimeter Wave Over Fiber Based 50G-Wavelength Division Multiplexed Passive Optical Network for 5G and Beyond Communications”, *Journal of Optical Communications*, juin 2023, doi: 10.1515/joc-2023–0091.
36. A. K. Banerjee, A. Sensoy, J. W. Goodell, and B. Mahapatra, “Impact of Media Hype and Fake News on Commodity Futures Prices: A Deep Learning Approach over the COVID-19 Period”, *Finance Research Letters*, vol. 59, p. 104658, janv. 2024, doi: 10.1016/j.frl.2023.104658.
37. J. Cui, Z. Wang, S.-B. Ho, and E. Cambria, “Survey on Sentiment Analysis: Evolution of Research Methods and Topics”, *Artificial Intelligence Review*, vol. 56, nº 8, pp. 8469–8510, août 2023, doi: 10.1007/s10462-022-10386-z.
38. S. Wu *and al.*, “A Comprehensive Exploration of Personalized Learning in Smart Education: From Student Modeling to Personalized Recommendations”. arXiv, 15 janvier 2024. doi: 10.48550/arXiv.2402.01666.

CHAPTER 12

5G and AI-based data fusion in intelligent networks

Hanane Lamaazi and Elarbi Badidi

1 INTRODUCTION

Artificial intelligence (AI) and fifth generation (5G) are the cutting edge of innovation, altering traditional communication networks and infrastructures to become more intelligent [1]. Intelligent networks (IN) present traditional networks converged to work with AI solutions toward more autonomous and intelligent decision-making while improving the quality of services and performances [2]. Incorporating AI models into critical and real-sensitive IN applications, such as 5G communication technology, guarantees good network performance quality, such as low latency, high speed, and reduced resource consumption. The main objective of 5G technology is to increase the current data rate by more than 10 Gbit/s and increase the available bandwidth using higher frequencies [3].

5G can support high-speed human interactions, especially when integrated with AI systems, the Internet of Things (IoT), cloud/edge computing [4], and software-defined networks (SDN). This allows INs to enhance performance, reliability, and efficiency and enables high-speed internet access and bandwidth-intensive applications. INs can support a huge number of connected devices simultaneously. These devices generate huge amounts

DOI: 10.1201/9781003514336-12

of data with different characteristics, sizes, and quality depending on the devices' nature, configuration parameters, and control capabilities [5]. Processing data involves extracting valuable features from raw data that provide a logical understanding of natural or artificial objects. Defining efficient data processing is a major challenge for many domains, especially data collected from heterogeneous sources, redundant and incomplete data, and time-series data [6]. To address these challenges, data fusion methods have been developed to aggregate and combine certain data to obtain the ground truth about a specific phenomenon regardless of the data being noisy, inconsistent, or inaccurate. Different types of data fusion can be used to process the data and can be simple to implement, while others require complex designs.

This chapter introduces the concept of INs, including the basics of 5G and AI technologies. It comprehensively analyzes the data fusion concept, including the different application levels, types of data fusion, and the most used techniques and algorithms to apply it.

2 5G AND AI IN INTELLIGENT NETWORKS

2.1 5G and beyond

5G mobile technology is a recent generation of mobile networks that support emerging applications and services. 5G delivers significantly advanced mobile broadband, ultrareliable low-latency communication, and massive machine-type communication [7]. Unlike the previous cellular networks, 5G allows an increase in the data rate and makes it available anytime and anywhere. It enhanced human interaction through virtual and augmented reality, telepresence, and free-viewpoint video. Hence, 5G can be deployed for various applications, including healthcare, entertainment, automotive, IoT, and industrial IoT applications that require ultralow latency for higher performance [8]. Integrating 5G technology with other technologies increases the number of connected devices that generate a huge amount of multisource data [7]. Thus, developing intelligent models to fuse the data by integrating AI models is necessary to overcome the challenges of multidimensionality, inconsistency, or incomplete data.

2.2 AI fundamentals

The concept behind AI is to enable machines to mimic human intelligence and problem-solving skills. It covers many methods, including rule-based systems and neural networks, and it is used in many industries, including smart cities, crowdsensing, autonomous vehicles, finance, healthcare, etc. The primary focus of the AI subfield of machine learning is developing methods and models that empower computers to learn from data and autonomously make predictions and decisions. AI relies on three main aspects: (1) learning, which involves the building of an algorithm that allows the computer to reinforce its learning through various experiments and tests; (2) reasoning, which entails making decisions based on logical studies and contextual modeling; and (3) perception, it incorporates the interaction with the environment through the use of multisensors as input to collect and sense the data [1].

3 DATA FUSION FUNDAMENTALS

Data fusion's primary application was in the military domain, and its first mathematical theories were introduced in the 1960s. In the 1970s and 1980s, data fusion was extended to be implemented in other domains and disciplines, aiming to promote the concept by developing projects, organizing conferences, and performing research on the data fusion concept [9]. The term "data fusion" is used in parallel with other terms, such as "sensors fusion" and "information fusion," which are commonly used to describe the process of aggregation and combining multisource information to extract more useful and accurate data. Data fusion improves the decision-making about the system's operations and state. Applied in INs, data fusion allows gathering different types of data and converting it into knowledge-based details to understand and explain the surrounding environment and make more reliable decisions [9]. The data fusion process can be implemented based on various criteria, such as the type, level, and chosen techniques. Making an optimal choice can help develop effective and intelligent solutions for better decision-making. In this section, different classification of data fusion is presented based on the level, type, and techniques presented as follows:

3.1 Levels of data fusion

Providing accurate and consistent information is crucial for several applications, such as healthcare, driver behavior assessment, and crash detection. These applications rely on integrating multiple sources to collect data. This data has different characteristics and is affected by the deployed sensors. Using data fusion techniques helps to enhance the quality of sensed data efficiently and with high accuracy. Data fusion techniques can be applied on different levels, from raw data integration to decision-making,

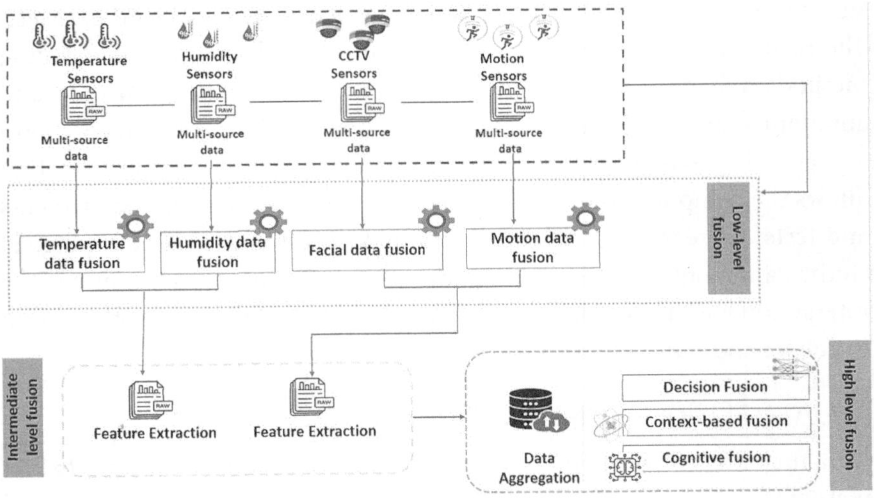

FIGURE 12.1 Systematic illustration of data fusion levels.

enhancing INs' capabilities and performances [10]. Data fusion levels can be presented as illustrated in Figure 12.1 and described as follows:

- **Low-Level (Sensor-Level) Fusion**: This level combines raw data from multiple sensors to produce a refined dataset. It aims to enhance the accuracy and reliability of the information collected from individual sensors. It can be applied using different techniques such as signal processing (e.g., filtering, noise reduction, etc.), data alignment (e.g., data synchronization), and statistical methods (e.g., Kalman filtering or Bayesian estimation) to improve the raw data quality, temporal consistency, and sensor readings. It can be used in autonomous vehicles, where data is collected from sensors such as radar, LiDAR, and cameras, or in healthcare, where various biosensor readings are merged to monitor vital signs accurately.
- **Intermediate-Level (Feature-Level) Fusion**: At this level, features are extracted from raw data collected from different sources and then fused to provide a more informative representation. It can be applied using feature extraction, pattern recognition, and dimensionality reduction techniques to identify patterns and reduce the complexity of features extracted from different sources. This fusion level is generally used in applications requiring high-quality images or monitoring machinery health, such as image processing and industrial automation.

- **High-Level (Decision-Level) Fusion**: According to the data inputs, the systems made fused decisions to reach a more reliable final decision. It involves techniques such as ensemble methods, rule-based systems, and voting schemes that help to aggregate decisions and improve their accuracy. This fusion level can be deployed in critical applications such as medical diagnosis, where merging multiple diagnoses leads to a better conclusion and then helps to provide optimal remediation.
- **Context-Level Fusion**: In this advanced level, the information is fused based on its context, including temporal, spatial, or situational contexts, using adaptive filtering or contextual modeling for better data interpretation. This fusion level is used on applications that rely on contextual data, such as smart cities, where a combination of, for instance, weather conditions, time, duration, and historical traffic patterns help in traffic management optimization.
- **Fusion at the Cognitive Level**: It is considered the highest fusion level dedicated to complex scenarios where the fusion of data, reasoning, and knowledge are required. The fusion process emulates human cognitive processes such as new learning, problem-solving, and interpretation. Cognitive computing, inference engines, and knowledge representation techniques can be used to fuse the data at the cognitive level and make more informed decisions. This fusion level is deployed in applications that use situational awareness systems, such as defense domain or advanced robotics, to promote strategic or autonomous decisions.

3.2 Types of data fusion

Data fusion can be categorized into different types based on how data from multiple sources are combined to achieve a more accurate and comprehensive understanding (see Table 12.1). Here are the primary types of data fusion:

- **Complementary Data Fusion**: Several sources can be deployed for specific phenomena to sense nonredundant and unique information, which, when combined, leads to more reliable and accurate insights. For example, in the healthcare domain, collecting heart rate and blood tests can help to determine the body's condition or identify disease symptoms [9].

TABLE 12.1 Overview of the various types of data fusion

			Data Source			Data Type	
			Sensors		Satellite	Historical	Real-time
Type of Data Fusion	**Techniques**	**Applications**	**Active**	**Passive**			
Complementary	Feature-level fusion	-Remote sensing	✓				✓
	Contextual integration	-Healthcare		✓		✓	
Redundant	Statistical methods (e.g., averaging, Bayesian inference)	-Navigation systems	✓	✓		✓	✓
	Signal processing (Kalman filtering)	-Sensor networks		✓			✓
Cooperative	Collaborative filtering	-Autonomous driving	✓				✓
	Multi-agent systems	-Collaborative robotics	✓				✓
Static	Data aggregation	-GIS (Geographic Information System)			✓		✓
	Pattern recognition	-Business intelligence				✓	
Dynamic	Real-time processing	-Autonomous vehicles					✓
	Adaptive algorithms	-Financial markets					✓
Hierarchical	Multi-level fusion	-Military applications	✓	✓		✓	✓
	Layered architectures	-Environmental monitoring		✓			✓
Multisensor	Sensor fusion algorithms (e.g., Kalman filters, neural networks)	-Robotics	✓				✓
	Data alignment and registration	-Weather forecasting			✓		✓

- **Redundant Data Fusion**: This type involves gathering data from various sources that provide the same information. The redundant data is fused to increase the reliability and accuracy of the information by minimizing noise and uncertainty. This type of fusion is used in accuracy-sensitive applications such as navigation systems, where combining GPS information from various satellites helps to improve location accuracy [9].
- **Cooperative Data Fusion**: In this type of data fusion, the information is actively extracted from different sources that collaborate and interact to provide more comprehensive and detailed information. For example, this fusion type can be used in autonomous driving, where multiple vehicles exchange sensor data to improve contextual awareness and safety significantly [9].
- **Static Data Fusion**: This type of fusion is used for applications that rely on data that does not frequently change or remains the same over time, such as demographic data, historical sales, and product descriptions. It can also deal with geographic information systems (GIS) that capture, store, and process data generated from coordinates on Earth's surface, such as latitude and longitude [11].
- **Dynamic Data Fusion**: Unlike static fusion, dynamic data fusion involves real-time data that frequently changes over time and is collected continuously from multiple sources. This type of fusion can be used in real-time applications such as autonomous vehicles, traffic management, stock prices, trading decisions, and crowdsensing [11, 12].
- **Hierarchical Data Fusion**: Hierarchical, layered, or multi-level fusion is used for applications requiring various processes to make a final decision, starting from raw data to the decision fusion level. For example, developing a new tactic or strategy in military applications can be made only if multiple data at different levels are aggregated and processed.
- **Multisensor Data Fusion**: This type aggregates data from multiple sensors capturing the same or different phenomena to enhance the overall understanding and ensure the consistency of the sensed data.

3.3 Data fusion techniques

Another classification of data fusion is based on different techniques selected based on their deployment context (see Table 12.2). Here are some key data fusion techniques used in INs:

TABLE 12.2 Overview of the various data fusion techniques

Technique	Key Features	Shortcomings					Complexity	Scalability	Robustness to Noise
		Computation Complexity	Resource Consumption	High cost	Parameter Selection	Others			
Kalman filtering	-Optimal for linear systems, -Real-time processing					✓	Moderate	High	High
Particle filtering	-Handles nonlinearities, -Monte Carlo simulations	✓					High	Moderate to low	Very high
Bayesian inference	-Probabilistic, -Incorporates prior knowledge	✓					High	Moderate	High
Maximum likelihood estimation (MLE)	-Parameter estimation, -Likelihood maximization				✓		Low	High	Moderate
Convolutional neural networks	-Feature extraction, -Hierarchical learning		✓				High	High	Moderate to high
Recurrent neural networks (RNNs)	-Temporal patterns, -Sequence learning					✓	High	High	Moderate to high
Voting schemes	-Majority/weighted voting, -Ensemble methods					✓	Low	High	High
Dempster–Shafer theory	-Belief functions, -Uncertainty management	✓					High	Moderate	High

(*Continued*)

TABLE 12.2 (*Continued*) Overview of the various data fusion techniques

		Shortcomings							
Technique	**Key Features**	**Computation Complexity**	**Resource Consumption**	**High cost**	**Parameter Selection**	**Others**	**Complexity**	**Scalability**	**Robustness to Noise**
Principal component analysis (PCA)	-Reduces dimensionality, -Retains variance					✓	Moderate	High	Moderate
Canonical correlation analysis (CCA)	-Correlation measurement, -Feature extraction					✓	Moderate	High	Moderate
Contextual modeling	-Context-aware					✓	High	High	High
Adaptive filtering	-Real-time adaptation, -Dynamic parameter tuning	✓					Moderate	High	High
Hybrid approaches	-Integrated methods	✓		✓			High	High	High

- **Kalman Filtering:** It is an optimal generic algorithm that uses noisy or inaccurate measurements to estimate and predict the state of dynamic systems with high accuracy. It is widely used in applications that involve continuous measurements, such as navigation, control, and target tracking [9, 13].
- **Particle Filtering:** It is a nonlinear filtering algorithm that implements a recursive Bayesian filter by Monte Carlo simulations to solve the filtering problem for nonlinear state–space systems. Basically, it is used to estimate the posterior density probability of the system by using a collection of discrete weighted samples, known as particles, that describe the probability assessments of the target state. It is generally used in applications such as Bayesian statistical inference and signal processing [13].
- **Bayesian Inference:** It is a statistical inference method that relies on observed data and prior knowledge to update beliefs and model uncertainty. It considers the probability distributions, possible outcomes, and existing knowledge to provide more solid decision-making and accurate predictions. It is mostly used in applications such as medical diagnosis, vision recognition, and time-sensitive decision systems for classification, clustering, regression, and reinforcement learning [14].
- **Maximum Likelihood Estimation:** It is a method of estimating more robust probability distribution parameters based on observed data to achieve maximum likelihood function. It is extensively used in various problems, such as world economy modeling, wireless communication networks, and population genetics [15].
- **Convolutional Neural Networks (CNNs):** CNNs are the most significant class of deep neural networks commonly used for natural language processing and computer vision. It involves extracting features or patterns by fusing input data with a convolution matrix. The CNN architecture consists of artificial neurons, kernels acting as receptors interacting with various features, and activation functions that simulate the transition of neural signals to the next neuron when exceeding the threshold. It is used for applications relying on image and video data, such as autonomous driving, vehicular surveillance, and traffic control [16, 17].

- **Recurrent Neural Networks (RNNs):** RNNs rely on updating and repeating their internal state due to their dynamic temporal behavior, allowing them to recognize a pattern in sequences of random length. RNN can be used to predict values or labels for every step in processing sequential data (e.g., detecting anomalies in time series) or when a sequence is terminated (e.g., music gender classification) [18].
- **Voting Schemes:** The voting process is an essential operation for ultrareliable systems relying on multichannel computation. It allows the fusion of data originating from multiple sources that provide incomplete, erroneous, or missing outputs. The voting schemes can be hardware or software and depend on the amount of data and the voting frequency. High-level voting requires software voting schemes to handle complex computations, while hardware voters can be deployed for low-level ones. A well-designed voting algorithm is more important than the implementation since it involves the integration of various requirements and constraints that consider the context of the input data object, the votes (weights), and the output data objects. It is widely used with ensemble learning techniques, particularly intrusion detection systems requiring high classification accuracy [19].
- **Dempster–Shafer (D–S) Theory:** It refers to the theory of evidence and presents a more general version of the Bayesian theory of subjective probability. It proliferates uncertainty values and consequently indicates the certainty of inferences [20]. It presents a comprehensive framework for reasoning about uncertainty and can connect with other frameworks such as probability theory, possibility theory, and imprecise probability theory. It extracts beliefs from the subjective probability of a particular problem and applies Dempster's rule to aggregate beliefs when new evidence is accumulated [20]. D-S theory is used in several applications such as risk analysis, target tracking, and fault diagnosis [21].
- **Principal Component Analysis (PCA):** PCA is mostly used to reduce the dimensionality of voluminous data while retaining most information and variability. Exploring and visualizing small datasets is much faster and less complicated for machine learning algorithms [22]. It can be used for highly correlated data, data with

multiple variables, or data with multiple iterations per variable. For example, PCA can be deployed in facial recognition and signal processing domains for data compression, noise reduction, and feature extraction [23].

- **Canonical Correlation Analysis (CCA):** CCA is a data analysis technique that measures the relationship between two multivariate sets of variables (canonical variates) to maximize the correlation (canonical correlations) between low-dimensional feature space and projected data. It relies on solving a singular value decomposition problem that entails covariance matrices of two multivariate datasets to find the pairs of canonical projection vectors. CCA can be used in different fields such as image processing and machine learning, social sciences, and biology [24].
- **Contextual modeling:** It allows the building of representations and models that incorporate surrounding conditions and the context of particular information, such as time, location, and user activities. The main concept of contextual fusion is that the interpretation and the meaning of information do not characterize the focal element of interest but may vary according to the situation in which they occur [25]. It is widely used in different domains, such as quantum physics [26] and intelligence analysis for counterinsurgency [27].
- **Adaptive filtering:** It involves techniques such as Kalman, particle, and adaptive Bayesian filters that dynamically adjust their parameters based on the context or changing environment to maximize reliability and improve the accuracy of fused information. It can be used for object tracking [28], target localization [29], and environmental monitoring [30].
- **Hybrid approaches:** It combines multiple fusion techniques such as sensor fusion, context-aware, or machine learning techniques to leverage and benefit from their strengths to obtain the desired result.

4 CONCLUSION

5G and AI-based data fusion techniques in INs have improved decision-making in different domains, such as smart cities, healthcare, and industry. They leverage a combination of statistical methods, signal processing, machine learning, deep learning, feature-level fusion, decision-level fusion, context-aware fusion, and hybrid approaches to create robust and efficient systems. An optimal choice of technique helps develop more efficient solutions that enable INs to handle dynamic and complex environments, address data imperfections and inconsistencies, improve decision-making, and support time and data-sensitive applications across various domains.

REFERENCES

1. H. Lamaazi and E. Mathew, "Comprehensive comparative analysis of artificial intelligence, machine learning, and deep learning," in *Artificial Intelligence for Intelligent Systems*, Inam Ullah Khan, Mariya Ouaissa, Mariyam Ouaissa, Muhammad Fayaz, Rehmat Ullah, Eds. CRC Press, Florida, USA, 2025, pp. 51–68.
2. C. Wang, J. Chen, Y. Yang, X. Ma, and J. Liu, "Poisoning attacks and countermeasures in intelligent networks: Status quo and prospects," *Digital Communications and Networks*, vol. 8, no. 2, pp. 225–234, 2022.
3. G. Ancans, A. Stafecka, V. Bobrovs, A. Ancans, and J. Caiko, "Analysis of characteristics and requirements for 5G mobile communication systems," *Latvian Journal of Physics and Technical Sciences*, vol. 54, no. 4, pp. 69–78, 2017.
4. H. Lamaazi, "Cyber security for edge/fog computing applications," in *Cyber Security for Next-Generation Computing Technologies*, Inam Ullah Khan, Mariya Ouaissa, Mariyam Ouaissa, Zakaria Abou El Houda, Muhammad Fazal Ijaz, Eds. CRC Press, Florida, USA, 2024, pp. 177–189.
5. T. Man, V. Y. Osipov, N. Zhukova, A. Subbotin, and D. I. Ignatov, "Neural networks for intelligent multilevel control of artificial and natural objects based on data fusion: A survey," *Information Fusion*, vol. 110, p. 102427, 2024.
6. F. Kibrete, D. E. Woldemichael, and H. S. Gebremedhen, "Multi-sensor data fusion in intelligent fault diagnosis of rotating machines: A comprehensive review," *Measurement*, vol. 232, p. 114658, 2024.
7. S. Hakak, T. R. Gadekallu, P. K. R. Maddikunta, S. P. Ramu, M. Parimala, C. De Alwis, and M. Liyanage, "Autonomous vehicles in 5G and beyond: A survey," *Vehicular Communications*, vol. 39, p. 100551, 2023.
8. V. S. Pana, O. P. Babalola, and V. Balyan, "5G radio access networks: A survey," *Array*, vol. 14, p. 100170, 2022.
9. S. Hassani, U. Dackermann, M. Mousavi, and J. Li, "A systematic review of data fusion techniques for optimized structural health monitoring," *Information Fusion*, vol. 103, p. 102136, 2023.
10. N. S. Dasappa, N. Somu *et al.*, "Multi-sensor data fusion framework for energy optimization in smart homes," *Renewable and Sustainable Energy Reviews*, vol. 193, p. 114235, 2024.
11. A. Fang and Y. Li, "Dynamic and static fusion mechanisms of infrared and visible images," *Pattern Recognition*, vol. 155, p. 110689, 2024.

12. E. Blasch, Y. Al-Nashif, and S. Hariri, "Static versus dynamic data information fusion analysis using DDDAS for cyber security trust," *Procedia Computer Science*, vol. 29, pp. 1299–1313, 2014.
13. Y. Bai, B. Zhang, N. Xu, J. Zhou, J. Shi, and Z. Diao, "Vision-based navigation and guidance for agricultural autonomous vehicles and robots: A review," *Computers and Electronics in Agriculture*, vol. 205, p. 107584, 2023.
14. J. P. Bharadiya, "A review of Bayesian machine learning principles, methods, and applications," *International Journal of Innovative Science and Research Technology*, vol. 8, no. 5, pp. 2033–2038, 2023.
15. C. Kosiol, L. Bofkin, and S. Whelan, "Phylogenetics by likelihood: Evolutionary modeling as a tool for understanding the genome," *Journal of Biomedical Informatics*, vol. 39, no. 1, pp. 51–61, 2006, Phylogenetic Inferencing: Beyond Biology. [Online]. Available: https://www.sciencedirect.com/science/article/pii/S1532046405000766
16. Z. Li, F. Liu, W. Yang, S. Peng, and J. Zhou, "A survey of convolutional neural networks: Analysis, applications, and prospects," *IEEE Transactions on Neural Networks and Learning Systems*, vol. 33, no. 12, pp. 6999–7019, 2021.
17. alooba, Convolution Matrices, access 2024. Available: https://www.alooba.com/skills/concepts/deep-learning/convolution-matrices/.
18. A. Tsantekidis, N. Passalis, and A. Tefas, "Chapter 5-recurrent neural networks," in *Deep Learning for Robot Perception and Cognition*, A. Iosifidis and A. Tefas, Eds. Academic Press, Cambridge, USA, 2022, pp. 101–115. [Online]. Available: https://www.sciencedirect.com/science/article/pii/B9780323857871000105
19. B. Parhami, "A taxonomy of voting schemes for data fusion and dependable computation," *Reliability Engineering System Safety*, vol. 52, no. 2, pp. 139–151, 1996. [Online]. Available: https://www.sciencedirect.com/science/article/pii/0951832096000129
20. J. Ye, S. Dobson, and S. McKeever, "Situation identification techniques in pervasive computing: A review," *Pervasive and Mobile Computing*, vol. 8, no. 1, pp. 36–66, 2012.
21. K. Zhao, L. Li, Z. Chen, R. Sun, G. Yuan, and J. Li, "A survey: Optimization and applications of evidence fusion algorithm based on Dempster–Shafer theory," *Applied Soft Computing*, vol. 124, p. 109075, 2022.
22. Z. Jaadi, A Step-by-Step Explanation of Principal Component Analysis (PCA), Feb 23, 2024. Available: https://builtin.com/data-science/step-step-explanation-principal-component-analysis
23. F. Kherif and A. Latypova, "Chapter 12-principal component analysis," in *Machine Learning*, A. Mechelli and S. Vieira, Eds. Academic Press, 2020, pp. 209–225. [Online]. Available: https://www.sciencedirect.com/science/article/pii/B9780128157398000122
24. L. Luo, W. Wang, S. Bao, X. Peng, and Y. Peng, "Robust and sparse canonical correlation analysis for fault detection and diagnosis using training data with outliers," *Expert Systems with Applications*, vol. 236, p. 121434, 2024. [Online]. Available: https://www.sciencedirect.com/science/article/pii/S095741742301936X

25. L. Snidaro, J. García, and J. Llinas, "Context-based information fusion: A survey and discussion," *Information Fusion*, vol. 25, pp. 16–31, 2015. [Online]. Available: https://www.sciencedirect.com/science/article/pii/S1566253515000032
26. P. Bruza, L. Fell, P. Hoyte, S. Dehdashti, A. Obeid, A. Gibson, and C. Moreira, "Contextuality and context-sensitivity in probabilistic models of cognition," *Cognitive Psychology*, vol. 140, p. 101529, 2023. [Online]. Available: https://www.sciencedirect.com/science/article/pii/S0010028522000652
27. M. P. Jenkins, G. A. Gross, A. M. Bisantz, and R. Nagi, "Towards context aware data fusion: Modeling and integration of situationally qualified human observations to manage uncertainty in a hard+soft fusion process," *Information Fusion*, vol. 21, pp. 130–144, 2015. [Online]. Available: https://www.sciencedirect.com/science/article/pii/S1566253513000675
28. J. Xiao, R. Stolkin, M. Oussalah, and A. Leonardis, "Continuously adaptive data fusion and model relearning for particle filter tracking with multiple features," *IEEE Sensors Journal*, vol. 16, no. 8, pp. 2639–2649, 2016.
29. Z. Guo, F. Zhao, Y. Sun, X. Chen, and R. Wu, "Research on target localization in airborne electro-optical stabilized platforms based on adaptive extended Kalman filtering," *Measurement*, vol. 234, p. 114794, 2024.
30. H. Li, M. Li, K. Zhan, P. Guo, X. Liu, X. Yang, and Z. Ma, "Study on application of multi-source data fusion method in environmental control of enclosed layer house," in *2023 IEEE 6th Information Technology, Networking, Electronic and Automation Control Conference (ITNEC)*, Chongqing, China, vol. 6. IEEE, 2023, pp. 1248–1252.

Index

AI 1, 18–21, 25–30, 33–34, 68, 80, 101, 112, 127, 158, 170, 187–188, 200–203, 205–206, 208–210, 212–213
antenna 3, 10, 37, 42, 74, 116–122, 124–125, 128–129, 132, 178
AR 37, 75–76, 84, 134–135, 174, 176–177, 179, 181–182, 193–195, 224

blockchain 34, 37, 54–59, 61–70

context-aware 211, 228–229

data fusion 218–221, 223, 228
DL 97, 101–109, 111–114, 132, 137, 199–203, 206–210, 212–213

edge computing 37, 114, 179, 194, 217
energy consumption 12, 15–17, 21, 25, 37, 76–77, 141–148, 162, 211

federated learning 37, 77, 79, 113, 133
5G 2–5, 7–12, 19, 25, 27, 32–33, 54, 58–61, 68–70, 90, 92, 96–100, 118, 121, 128–133, 156–158, 161–165, 174, 177–179, 194, 205–206, 209–212, 228

intelligent networks 17, 217–218
IoT 1, 9–11, 19, 55, 58, 68, 70, 75, 77–78, 81, 85–87, 90, 92, 98–99, 111, 118, 167, 217–218

MIMO 3, 12, 37, 74, 96, 100, 111, 116–120, 130–133, 136, 156, 178–179
ML 18–20, 22–25, 32–34, 77–81, 87–88, 92, 101, 111, 119–120, 125, 130–134, 136–137, 141–143, 152, 157–158, 160, 170–171, 201–202, 204, 207–212
M2M 40, 45, 49–50

network management 28, 36–37, 58, 64, 70, 76–78, 82, 85, 88–90, 92, 129, 134, 159, 205
NG-RAN 5–6
NOMA 74, 95–114, 123, 125–126, 136

power consumption 12, 16, 25, 111, 129, 140–145, 150–152, 159, 187

QoS 4–5, 7–8, 18, 24–33, 36, 40–41, 43–44, 49, 55, 61, 66, 76, 83, 86, 92, 102, 105, 114, 157, 159, 161–166, 169–170, 209

RACH 41, 45, 48, 52
reinforcement learning 19, 103, 105, 109, 132, 143, 157, 160–161, 164, 211–212
resource allocation 15–19, 21, 24–27, 29–36, 38, 44, 48, 50–52, 60–61, 65, 68, 77, 82, 85–86, 90, 97, 103, 107–111, 132, 134, 157, 159, 162–163, 165–166, 169
RRM 43, 49–51, 54–56, 59–70, 209

scheduling 5, 40–44, 49–52, 86, 105, 109, 143–146, 156–170
signal detection 111–112, 124, 126–127, 136

6G 2, 10–12, 36–37, 77, 97, 133–136, 167–168, 209–210
smart contract 56–58, 62, 64–66, 68, 70
spectrum 3, 10, 12, 17, 41, 49–50, 58, 60, 62–67, 74–75, 84, 96, 98–99, 119, 121, 131–132, 136, 156, 159, 180
VR 37, 75–77, 134–135, 174, 176–177, 181–182, 193–194

wireless communication 11, 54, 59, 95, 102, 111–112, 116–123, 127–128, 131–134, 166, 209, 226